Microsoft®

flight
simulator
as a **training aid**

Microsoft® flight simulator
as a training aid
a guide for pilots, instructors, and virtual aviators

bruce williams

Aviation Supplies & Academics, Inc.
Newcastle, Washington

Microsoft® Flight Simulator as a Training Aid: A Guide for Pilots, Instructors, and Virtual Aviators
with companion CD
by Bruce Williams

Aviation Supplies & Academics, Inc.
7005 132nd Place SE • Newcastle, WA 98059
(425) 235-1500 • email asa@asa2fly.com
Internet: www.asa2fly.com

None of the material in this manual supersedes any operational documents or procedures issued by the Federal Aviation Administration, aircraft and avionics manufacturers, flight schools, or the operators of aircraft.

Microsoft® Flight Simulator screen shots reprinted with permission from Microsoft Corporation.

Products photos by Logitech, Thrustmaster, Saitek, CH Products, Cub Crafters, and TRC Development b.v., used by permission; photos provided courtesy Alteon Training L.L.C., Galvin Flying, Jacqueline Gauger of FlightSafety International Academy, Rob Howard, Jeremy Wilson, and USAF Air Force Links photos at www.af.mil/photos. Page ix and back cover author photo courtesy Pat DuLaney. Page 36 bottom photo courtesy Jeremy Wilson; page 103 photo courtesy Cub Crafters, Inc. Cover photo of Cessna 172 courtesy Jeremy Wilson.

Printed in the United States of America

09 08 07   9 8 7 6 5 4 3 2

ASA-MSFS

ISBN 1-56027-670-3
      978-1-56027-670-8

Library of Congress Cataloging-in-Publication Data:
Williams, Bruce
  Microsoft flight simulator as a training aid : a guide for pilots, instructors, and virtual aviators / Bruce Williams.
       p.   cm.
  Includes bibliographical references and index.
  ISBN-13: 978-1-56027-670-8 (pbk.)
  ISBN-10: 1-56027-670-3 (pbk.)
  1. Microsoft Flight simulator (Computer file) 2. Flight training—Data processing. 3. Flight simulators—Computer programs. 4. Computer flight games—Computer programs. I. Title.
  TL712.8.W55 2006
  629.132'52028553—dc22
                              2006037919

12

contents

	foreword by Rod Machado	vii
	introduction	ix
one	about this book	1
two	using Flight Simulator as a training aid	15
three	best practices for using Flight Simulator	51
four	Flight Simulator essentials	67
five	advanced "training features" in Flight Simulator	89
six	differences between Flight Simulator 2004 and Flight Simulator X	97
seven	about the practice flights	103
eight	flying the aircraft used in the practice flights	117
nine	supplemental information and web links	141
ten	introduction to the VFR practice flights	151
eleven	sample briefings for VFR practice flights	179
twelve	introduction to the IFR practice flights	187
thirteen	sample briefings for IFR practice flights	217
fourteen	creating your own practice flights	227
	index	239

foreword

by Rod Machado

When I first met Bruce Williams, we were at an aviation convention and struck up a conversation on flight simulation. We've been friends ever since. In the flying business, Bruce is a rare bird. On one hand, he's an accomplished flight training expert; on the other, he probably knows more about using flight simulation as a training tool than anyone south of the North Pole. This unusual mix of practical cockpit experience, flight simulation knowledge, and a full spectrum of flight training skills makes Bruce one of the leading educators in the aviation simulation business. When he talks, pilots throttle back and listen. Now his unique insights are available for everyone to read.

Having created and voiced the flying lessons and ground school in the last several editions of Flight Simulator, I'm very familiar with the product and its use in flight training. I also know good training materials when I see them. *Microsoft Flight Simulator as a Training Aid: A Guide for Pilots, Instructors and Virtual Aviators* impressed me with both the depth and breadth of its contents from the first time I read it.

In this book, Bruce takes Flight Simulator to its highest level as a flight training tool. This is a guide for anyone wanting to maximize their real or virtual cockpit experience. And why would anyone not want to do that? It's hard to argue against learning more efficiently, with less stress, and having more money left in the bank at checkride time. By letting Bruce become the flight training middle man through his enlightening book, you'll find efficient, novel and fun ways to learn about flying.

Suppose you're a student pilot who wants to prepare for an upcoming cross country flight. How might Flight Simulator help you do it? Bruce suggests first making the actual cross country flight, leg for leg and turn for turn, in the simulated world. The real flight then produces the sensation of *deja-flew*, resulting in less stress, greater pleasure, and a more thorough learning experience.

Flight instruction also achieves a cutting edge level with many of Bruce's keen training suggestions. Did you know, for instance, that it's possible to make Flight Simulator movie files of specific flights or maneuvers and use them to demonstrate the skills you want a student to learn? Your students can manually insert these files into Flight Simulator on their PC, run the movies, and try before they fly. Endless repetitions at no cost provide a unique opportunity to acquire and hone a skill before getting into the plane.

Ground instructors can use Flight Simulator in the classroom to show the airplane and all its components in action, on a large screen. Imagine students watching a ground school instructor fly an actual ILS approach to minimums instead of having him or her describe the experience in flat, textbook language. Now *that's* an impressive training tool.

Bruce understands the idea that practice doesn't necessarily make our skills *perfect*, but it can make them *permanent*. This book's large assortment of practice flights and the guidance on how to best fly them make it easy for any pilot or instructor, real or virtual, to attain the highest level of flight proficiency they choose to seek.

While it's true that sometimes a cigar is just a cigar, Bruce disabuses us of the notion that our desktop PC simulator is just another simulator. It isn't. *Microsoft Flight Simulator as a Training Aid* is a title and a promise that shows you how to make flight simulation a real-world cockpit asset.

Rod Machado
San Clemente, CA
April, 2006

introduction

I grew up around airplanes. My father was a career pilot in the Air Force, and like many kids in flying families, I immersed myself in aviation from an early age. I assembled models, read books about airplanes and pilots, and, in those pre-PC days, built cockpits out of cardboard boxes, complete with Tinkertoy control yokes and "displays" powered by my imagination.

I started flying lessons while I was in high school, soloed in the early 1970s, and over the last 30-odd years, I've added a variety of pilot and flight instructor certificates and ratings to my wallet. In the mid-1980s, I edited *The Western Flyer*, the predecessor to *The General Aviation News*, and today I continue to write and speak about aviation. I'm also an active aerobatic pilot and flight instructor.

During the last half of my 15-year career at Microsoft, I worked as a technical editor and business development manager on six versions of *Microsoft Flight Simulator*, beginning with the last edition released for MS-DOS and finishing with *Microsoft Flight Simulator 2004: A Century of Flight*.

why this book?

Many students, pilots, instructors, and others in the aviation community have a hunch that PC-based simulations—and *Microsoft Flight Simulator* in particular—can be used as aids in flight training and to help pilots maintain proficiency. But when computers appear, confusion often ensues, and many people in the aviation community aren't sure how best to employ tools like Flight Simulator. Although many pilots and instructors have told me they have used Flight Simulator during their flying careers and even were inspired to pursue flight training by their early encounters with Flight Simulator, they often aren't sure how to get the most out of the experience of virtual flying today. I hear many questions on this topic when I present Flight Simulator to pilots and aviation enthusiasts at such events as EAA AirVenture, AOPA Expo, flight instructor clinics, pilot gatherings, and safety seminars.

Several years ago, I wrote an article, "Microsoft Flight Simulator as a Training Aid," for the Flight Simulator Learning Center and product Web site. It's a general introduction to the subject. This book expands and consolidates my answers to the common questions and, I hope, gives the aviation community specific, realistic advice about how best to use Flight Simulator as a training aid.

More importantly, to respond to many requests for a syllabus and useful tools that will help pilots and virtual aviators (Flight Simulator hobbyists) use Flight Simulator effectively, this book describes Practice Flights I've created to help owners of *Microsoft Flight Simulator* develop and improve specific flying skills.

This book and the accompanying Practice Flights emphasize instrument flying skills and procedures, but I've included several Practice Flights that primary students, with the guidance of their instructors, can use to learn about basic flying skills such as making sense of the instruments and controls in a typical light-plane cockpit, navigating with the VOR system, and the learning the fundamentals of attitude instrument flying.

And I hope that virtual aviators—the millions of aviation enthusiasts around the world who share a passion for flying but don't pilot real aircraft—will also find this book helpful as they learn and apply much of the same theory and skills that certificated pilots must master.

In short, my goal is to help everyone who loves flying—virtual aviators, students, pilots working on new ratings, and their instructors—get the most out of every hour enjoyed in the air or the virtual skies.

Finally, I want to thank my former colleagues on the *Microsoft Flight Simulator* team, veterans of many versions of Flight Simulator and pilots all, who provided valuable comments and corrections, especially Hal Bryan, Mike Lambert, and Mike Singer. As always, Rod Machado offered his unique perspective, encyclopedic knowledge, and vast experience to the project.

one

about
this
book

checklist

□ who should
  use this book

□ get the most
  from this book

□ website for
  this book

□ learning about
  Flight Simulator

□ joysticks, yokes,
  and throttles

□ what this
  book isn't

□ charts and
  references

□ safety note

I worked on six versions of *Microsoft Flight Simulator* at Microsoft, and as part of my duties, I attended innumerable aviation trade shows and fly-ins around the world. I watched countless people of all ages take to the virtual skies, and I gave scores of presentations on how students, pilots, and instructors can use Flight Simulator as a training aid. I also answered thousands of questions (actually, the same dozen or so common questions thousands of times) from folks curious about how they could best complement time in a real cockpit with hours "flying" Flight Simulator.

Students, pilots, and instructors often tell me how they have used Flight Simulator during their flying careers. Many say encounters with Flight Simulator even inspired them to pursue flight training. If you read aviation magazines, attend trade shows, and visit popular online aviation forums, doubtless you've run across similar stories and questions about using PC-based flight simulations.

Many aviators credit *Microsoft Flight Simulator* with obvious benefits related to practicing instrument approaches and exploring unfamiliar airports. But my encounters with pilots and instructors and time spent using Flight Simulator with my own students suggest that the aviation community isn't getting the most out of virtual flying.

All of these experiences have led to this book, which I hope provides general suggestions, specific advice, and practical tools you can use to make effective use of Flight Simulator, regardless of the type of flying you do.

Before getting into the details of how to use Flight Simulator, however, it's best to review some pre-flight checklists.

who should use this book

To begin, here's a short list of the people who can benefit from reading this book and from using the Practice Flights and other resources designed to work with it:

- Student pilots (pre-private pilot) who want to enhance book-learning and review specific concepts and skills.
- Certificated pilots hoping to complement their real-world flying with additional hours in the virtual skies, upgrade their navigation skills and learn about advanced aircraft and procedures.
- Instrument rating students looking for ways to add interactivity to their IFR theory studies, to preview lessons, and polish specific IFR flying skills such as the use of advanced avionics and instruments (e.g., HSI and RMI).
- Flight instructors looking for new teaching tools for ground school classes and pre-flight and post-flight briefings.
- Virtual aviators (Flight Simulator hobbyists) who want to learn more about real-world flying to enhance their enjoyment of virtual aviation.
- Kids and adults who want to prepare themselves with a little more knowledge before they begin formal flight instruction.
- Teachers using Flight Simulator in the classroom to complement aviation-related lessons or programs.

what you need to get the most out of this book

Although many of the recommendations described in this book could apply to other PC-based flight simulations and training devices, I assume you have *Microsoft Flight Simulator*, specifically either

- *Microsoft Flight Simulator 2004: A Century of Flight* (Version 9)
- *Microsoft Flight Simulator X* (Version 10). To fly the Practice Flights that use the Garmin G1000 "glass cockpit," you must have the deluxe edition of Flight Simulator X.

No add-on aircraft, additional scenery, or other enhancements are required to use the Practice Flights discussed in this book.

Microsoft Flight Simulator X was still in development as I wrote this book. The new version builds on previous releases, so most of the features described here appear in both editions, and in general they work the same way. However, there are important differences between the two versions, some of which I discuss in Chapter 6, "Differences Between Flight Simulator 2004 and Flight Simulator X."

> **note**
>
> To learn more about *Microsoft Flight Simulator*, including its features and detailed system requirements, visit the official *Microsoft Flight Simulator* website at **http://fsinsider.com**. You'll also find links to how-to articles, technical support, and other information that will help you get the most out of Flight Simulator. I especially recommend the article "A Real-World Pilot's Guide to Flight Simulator."

the website for this book

This book is in part an attempt to combine the familiar, portable, and easy-to-use medium of printed pages with the flexibility, timeliness, and interactivity of the Web and PC-based simulation.

In addition to the website addresses and footnotes you'll find throughout these pages, I have consolidated the resources associated with this book, plus links to background information and other sites that I find useful, on my website. Putting all the Web-based resources in one place makes it easier to keep the Practice Flights and Web links up-to-date. And you need to remember only one Web address: **www.BruceAir.com**.

The specific resources available on my Web pages change, but you'll always find information and links in these general categories:

- Resources related to *Microsoft Flight Simulator*, including information about where you can find add-on aircraft and other accessories.
- Expanded information about, and links to, resources associated with the topics in this book, including aviation references, training manuals and other learning resources, many of which are free to download.
- The complete set of Practice Flights for *Microsoft Flight Simulator* described later in this book.

learning to use Flight Simulator

Don't panic if you're new to *Microsoft Flight Simulator* or if you haven't spent much time with it lately. If you're a pilot, you're already familiar with aircraft instruments and controls, aviation terminology, and basic flying skills. If you are learning to fly, you are acquiring that knowledge with help from your instructor. Because Flight Simulator is a *simulation* that reproduces the experience of flying an aircraft (as closely as possible on a PC-based platform), most of your aviation knowledge will transfer directly to virtual flying. In fact, if you have some aviation knowledge and experience, you are ahead of most novice virtual aviators who must figure out how to interpret flight instruments, decipher charts, and master such arcane skills as VOR navigation.

In any event, you don't have to memorize lists of keyboard commands or use menus for most "flying" tasks. In fact, you don't even need to keep a keyboard handy after you start Flight Simulator and select the initial conditions for your flight. Only a few basic computer-related skills are necessary to use Flight Simulator effectively, and I discuss them in detail in Chapter 4, "Flight Simulator Essentials." Even if you are already familiar with Flight Simulator, you will find Chapter 4 worth a quick review. Where appropriate throughout the book, I include suggestions and specific tips about how best to use Flight Simulator in various situations.

The Cessna 172 instrument panel in Flight Simulator closely resembles the real thing.

Of course, this book does not attempt to explain all the details of using *Microsoft Flight Simulator*. The Learning Center is a web-like guide to Flight Simulator that was installed with your software. It contains hundreds of pages of information, videos, pictures, illustrations, and links that tell you all about the features in Flight Simulator. You will also find many helpful articles about using Flight Simulator on the official *Microsoft Flight Simulator* website and on the website associated with this book.

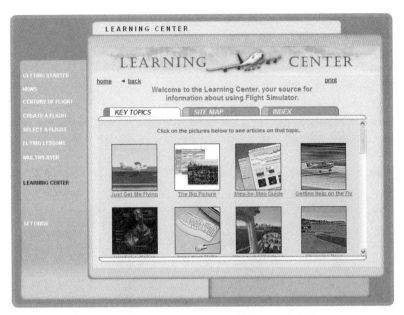

Microsoft Flight Simulator includes a Learning Center to help you get up to speed.

computer requirements

The short answer to the question about what kind of computer you need to enjoy *Microsoft Flight Simulator* is straightforward. If your computer is no more than 2–3 years old, it probably has the basic horsepower (i.e., processor speed and hard disk space) to run Flight Simulator—provided you have a good video (graphics) card and at least 512 Mb of system RAM.

Many budget systems sold today don't include high-performance graphics cards, so be sure to check that vital component, especially if your computer has "shared memory" or "integrated video," indicators that the system has a video chip built in to the motherboard instead of a separate graphics card with its own dedicated microprocessor and video memory. Fortunately, it is easy to find an inexpensive add-in video card that will work well with Flight Simulator.

 The website for this book—**www.BruceAir.com**—includes links to articles and other information about video cards and other computer-specific matters.

To learn more about configuring your system for maximum performance, see the following articles on the Flight Simulator website:

- "Optimizing Visuals and Performance"
- "Changing Display Settings"

All of my instructions for using Flight Simulator and the Practice Flights associated with this book assume you are running Windows XP or Windows Vista. If your computer can run Flight Simulator well, it is probably using one of those versions of Windows.

Of course, you also need access to the Internet to use the resources associated with this book and to download the Practice Flights. A broadband connection is recommended, but not required.

joysticks, yokes, throttles, rudder pedals, and other controls

Pilots often ask me if, to get value out of virtual flying, they need an expensive control yoke or joystick, rudder pedals, a throttle quadrant, and consoles to simulate the avionics stack and other aircraft controls. All of these items can make virtual flying more entertaining, but as I explain throughout this book, you don't need to re-create a full cockpit in your basement. The only accessories you need to use Flight Simulator effectively, and to apply the resources in this book, are a basic joystick and a mouse.

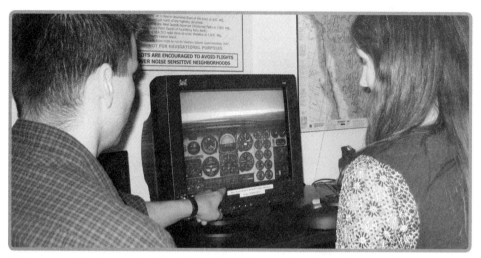

Minimum required setup for using Flight Simulator as a training aid

In fact, if you are an instructor and plan to use Flight Simulator as a training aid in the classroom and during briefings with students, you can run it on a late-model laptop with a dedicated graphics chip (many newer laptops have this capability). Joysticks and other peripherals manufactured in the last several years require a USB (Universal Serial Bus) connector, and they work fine with current laptops. I often give presentations about and demonstrations of Flight Simulator to pilot groups using only a laptop, mouse, and basic joystick.

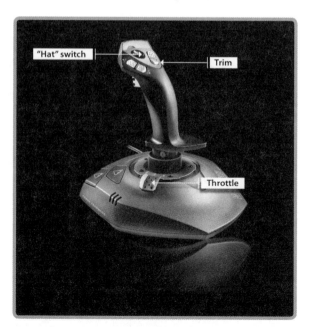

Typical computer joystick and controls

As I hope later chapters of this book make clear, other accessories are not necessary (or often even desirable) to make effective use of Flight Simulator as a training aid. I log most of my virtual flight time with a basic joystick. Like most joysticks available today, it has a built-in throttle, buttons to extend and retract the landing gear and flaps, a "hat" switch for looking around, and additional knobs, buttons, and levers for such functions as elevator trim and brakes. The joystick also twists to provide rudder control. Equally important, a joystick is simple to set up, compact, portable, and quickly put aside when I need my computer for terrestrial pursuits. As I point out in Chapter 4, "Flight Simulator Essentials," I use the mouse just as I would use my hand to adjust power, tune and select radios, configure the autopilot, set the heading bug, change VOR course settings, operate the GPS, and accomplish other tasks in the cockpit.

You can customize the controls on a joystick or yoke to suit your preferences.

If you don't like the default functions assigned to the various knobs and buttons on the joystick or yoke that you use with Flight Simulator, you can set up a custom configuration. For example, you might prefer to use the "hat switch" to adjust the elevator trim and reassign other buttons to switch views. The article "Using a Joystick" in the Flight Simulator Learning Center explains how to customize joystick assignments. You can also re-assign key combinations. For more information, see the article "Using the Keyboard" in the Learning Center.

Many manufacturers of joysticks and yokes have also created "profiles"—pre-configured assignments for the buttons and knobs on their devices—for *Microsoft Flight Simulator*. You can download these profiles from their websites.

At present, the leading manufacturers of joysticks for PCs are Logitech, Thrustmaster, Saitek, and CH Products. Microsoft no longer makes SideWinder joysticks for PCs. Each manufacturer posts the latest information about its products on its website.

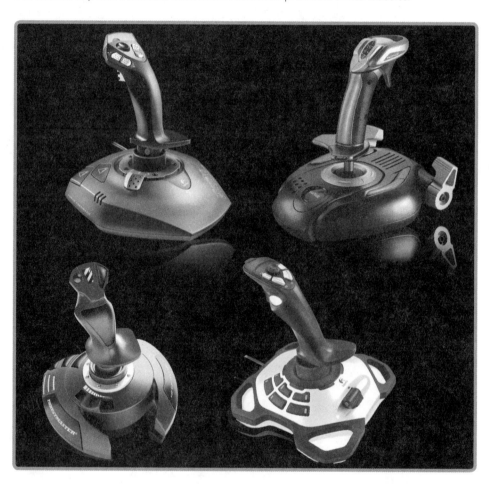

Of course, the typical general aviation airplane still has a yoke sticking out of the panel, not a joystick attached to the floor, and if you feel more comfortable flying with a yoke, you have a few options. The best device for most users currently available is the CH Products Flight Sim Yoke.

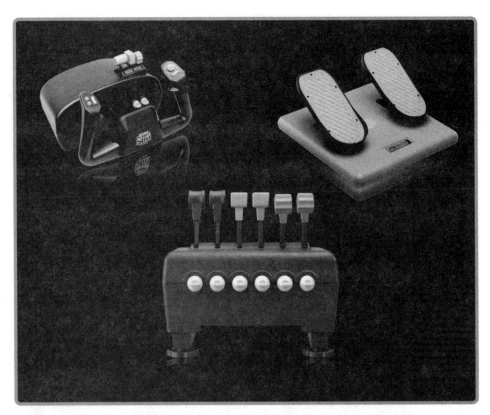

A yoke, rudder pedals, and throttle quadrant from CH Products.

> **note** You can find a link to a detailed review of the CH Products Flight Sim Yoke and more information about joysticks, components for virtual cockpits, and other accessories on the webpage for this book—**www.BruceAir.com**.

CH Products also makes a throttle quadrant that you can configure to control a piston single or twin, a turboprop, a twin-engine jet, or a four-engine jet transport.

A few manufacturers also make rudder pedals, but I have yet to find an affordable set that provides a realistic feel and control response. As I explain later, real airplanes are best for practicing slips, crosswind landings, and the finer points of flying taildraggers. The "Autorudder" feature in Flight Simulator can handle coordination chores for most routine flying, including the Practice Flights associated with this book.

Other options abound. If you do want to create a sophisticated home cockpit, you can find all the necessary components by shopping on the Web. Several vendors provide kits to build sophisticated instrument panels, avionics stacks, elaborate displays, and flight and engine controls.

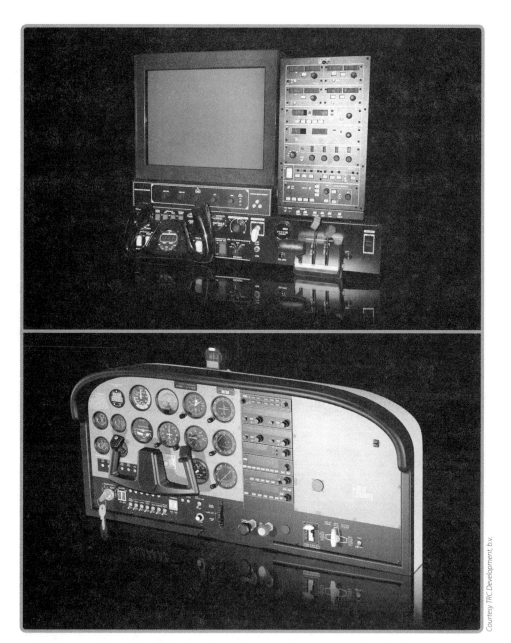

Courtesy TRC Development, b.v.

Several companies make sophisticated "cockpits" and components for use with Microsoft Flight Simulator.

My short answer to the question of which controls and accessories are best is to explore the offerings at any store or website that sells computer accessories (e.g., Best Buy, Circuit City, or Wal-Mart) and decide what best suits your preferences and budget. Prices for joysticks range from about $25 to around $70. Yoke-throttle-rudder-pedal combinations typically cost $100 to $200 per component. You can find links to information about many current products on the website for this book.

realistic expectations about realism

I have more to say about the "realism" provided by PC-based flight simulations in Chapter 2, "Flight Simulator as a Training Aid," but a few points are in order here.

Regardless of which gizmos you connect to your computer, don't expect too much from PC-based controllers. No one has figured out how to make realistic flight controls for PC simulations at prices that most home users are willing to pay. If you want yokes that reproduce varying control loads and precise transfer of inputs (especially in pitch), you have to spend many hundreds, if not thousands, of dollars for specialty components. Even then, you're likely to notice significant differences between a real airplane and the simulator.

This limitation doesn't apply only to Flight Simulator. I often fly an expensive, FAA-approved simulator at the flight school where I teach. It's a wonderful training device that can help pilots develop and refine flying skills as well as learn and practice specific instrument procedures. Although the simulator accurately reproduces the *performance* of the aircraft it emulates (C172 Skyhawk, BE-36 Bonanza, and PA-44 Seminole), it does not *feel* much like a real airplane. More importantly, as I explain in the next chapter, using Flight Simulator as the foundation for a painstakingly reproduced cockpit is not necessarily the most effective way to incorporate PC-based simulation into flight training.

what this book isn't

This book and the accompanying Practice Flights are not intended to provide all the background knowledge pilots and virtual aviators need to fly. Many excellent, thorough, and authoritative training handbooks, references, and online resources already cover that territory. Instead of producing yet another treatise about the fundamentals of flying, I have provided references to assets—many of them available for free download from the Web—that I find helpful and that I recommend to my students and fellow aviators. I provide descriptions of specific references later in this section, and you can find details about and links to these books, online courses, and other publications on the website for this book.

Of course, where I think I have something of value to add about a specific flying technique or procedure, I offer suggestions and recommendations, and each Practice Flight includes a briefing with specific references to appropriate background information and related topics.

You can find detailed information about the resources that I recommend to complement the Practice Flights in Chapter 9, "Supplemental Information and Web Links," but some are worth brief mention here.

FAA aviation handbooks

The FAA offers many complete, public-domain training handbooks in Adobe Reader format (.pdf) for download from its website. These books are the official references for many pilot certificates and ratings, and they are comprehensive, practical guides to the core knowledge every pilot should possess. See Chapter 9 for a list of specific titles referred to in the Briefings for the Practice Flights that accompany this book.

The best places to find the latest versions of these handbooks and other official references are the FAA's website and through the links on this book's website. The FAA includes new titles periodically, so check its website occasionally to see what has been added to the virtual bookshelf.

charts and other references

Charts, diagrams, and other references required to use specific Practice Flights are included on the CD that accompanies this book.

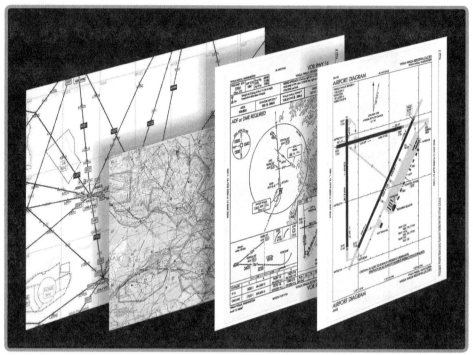

Examples of NACO charts

Jeppesen charts

If, like many IFR pilots, you prefer Jeppesen charts, you can learn all about them by downloading the free Chart Clinic Reprints available on the "Aviation Resources" section of the Jeppesen website. For more information on this series, see Chapter 9.

aviation books

If you prefer to have printed books handy as you learn, ASA (www.asa2fly.com) publishes reprints of the FAA training handbooks mentioned above and many excellent original titles. Of the latter, I specifically recommend:

- *FAR/AIM* (published annually)
- *The Complete Private Pilot* by Bob Gardner
- *The Complete Advanced Pilot* by Bob Gardner
- *Pilot's Manual Volume 1: Flight School*
- *Pilot's Manual Volume 2: Ground School*
- *Pilot's Manual Volume 3: Instrument Flying*

ASA titles are available from the ASA website (www.asa2fly.com) and at most pilot shops and bookstores.

FAA training handbooks

other useful references from the AOPA Air Safety Foundation

A wealth of practical information for pilots is also available free to everyone from the AOPA Air Safety Foundation Online Safety Center (www.aopa.org/asf/). Resources there include a variety of publications in PDF format, free online courses, and databases. I especially recommend the Safety Advisors and Safety Briefs available in the ASF Library. These colorful, clearly written booklets are offered primarily to enhance safety, but they are also excellent summaries of important knowledge and procedures, and they provide terrific introductions to, and reviews of, many operational topics.

AOPA has also created a special Web page for Flight Simulator aviators. That page provides links to articles from AOPA publications and other resources that are especially useful to virtual aviators.

important note

The Practice Flights described in this book and its associated website are not a substitute for instruction from an authorized flight instructor, and they do not cover all the knowledge and skills required of a pilot operating a real airplane in the real skies. They are intended to complement formal ground and flight instruction, to provide practice for real-world pilots and to help virtual aviators get more enjoyment from flying *Microsoft Flight Simulator*.

If you intend to use the Practice Flights during training for any pilot certificate or rating, while preparing for a flight review or instrument proficiency check, or for brushing up your real-world flying skills, please consult with an instructor to ensure you are developing good, safe flying habits and mastering the theory, practical knowledge, skills, and procedures according to the latest standards and requirements.

In addition, the Practice Flights—especially the IFR Practice Flights—focus on flying according to rules, procedures, and standards applicable in the U.S. at the time this book was published. Pilots obviously need to know, understand, apply, and adhere to the current regulations and procedures established by the authorities where they fly.

two

using Flight Simulator as a training aid

checklist

☐ deconstructing Flight Simulator

☐ obstacles to learning

☐ a "swiss army knife"

☐ interactive chair flying

☐ negative transfer

☐ learning environment

☐ flight models and controls

☐ flying my airplane

☐ self instruction

☐ instrument panel fixation

☐ IFR procedures in Flight Simulator

☐ learning curve

☐ learning, not logging

Aviation has come a long way from "needle, ball, and airspeed" and the "dit-dah, dah-dit" of the radio range. Yet outside of the NORAD-like classrooms at aviation-oriented universities and the computer-based training programs offered to professional pilots, most flight schools teach flying much as they always have done, even as 21st century versions of the Piper Cub emerge from the factory with electronic flight instruments and color moving maps driven by GPS receivers.

Most of us on the bottom rungs of the aviation-training ladder still instruct using the same tools and methods that were in use when we learned to fly, largely because independent flight schools haven't had access to cost-effective, easy-to-use versions of the devices and techniques employed by flight-training institutions. Experienced aviators agree that an airplane is a terrible place to introduce new concepts, but because many instructors think we still lack an effective substitute, the cockpit remains our primary interactive, real-time demonstration, practice, and evaluation tool.

With the advent of the personal computer, some pilots and instructors immediately recognized that it could be used for more than creating syllabi and PowerPoint presentations, especially because *Microsoft Flight Simulator* was one of the first widely available software titles that didn't just process words or tote-up spreadsheets. Pilots and instructors with experience using certified flight training devices and simulators naturally thought of *Microsoft Flight Simulator* as a less-expensive version of those familiar tools, and some flight schools and computer-savvy independent instructors have long included *Microsoft Flight Simulator* in their courses.

"Glass" panel in a Sport Cub LSA

The cockpit remains the classroom of choice for many flight schools.

Flight Simulator lab at the FlightSafety International Academy

These folks take PC-based simulation seriously. Many instructors insist proudly that they allow students to use Flight Simulator only if it's running on a computer equipped with realistic flight controls—at a minimum a yoke, rudder pedals, and throttles. They impose these requirements for good reasons, among them: to emphasize the seriousness of the task at hand, to instill good habits (and limit adverse side effects), and to reinforce real-world procedures.

Cockpit by Simkits

Flight Simulator can indeed become the core of an immersive virtual cockpit, especially if equipped with a yoke, throttles, and rudder pedals, and separate panels for avionics, flap levers, fuel selectors, and the like. In fact, *Microsoft Flight Simulator* has spawned an industry to supply such accessories. Ambitious hobbyists have created astonishingly detailed cockpits in their basements and garages.

A few manufacturers have also developed elaborate training devices driven by *Microsoft Flight Simulator*, some of which have been approved by the FAA and other aviation authorities.

Rob Howard photo

Home cockpit built by Rob Howard

Organizations like the U.S. Navy and the Flight Safety International Academy include *Microsoft Flight Simulator* as an informal part of their flight training programs. These schools typically set up dedicated Flight Simulator workstations and cockpit pods, which they use to complement the "real simulator" and flight time included in their training programs. Students have access to the systems in a computer lab where they can practice on their own time. Instructors trained to use the specialized equipment are available to assist students as they prepare for and review flight lessons.

> **note**
>
> The website for this book includes links to articles and other information that discuss the U.S. Navy's Microsim program and other uses of *Microsoft Flight Simulator* in real-world flight training.

deconstructing Flight Simulator

All of these applications of *Microsoft Flight Simulator* have their place. The Navy and other institutional users have documented many benefits of the "Flight Simulator lab" approach, including the ability to practice scenarios that are hazardous or difficult to re-create in the air, and, perhaps more importantly, to deliver quality training more efficiently than is possible in a real airplane.

Yet restricting Flight Simulator's role to re-creating flights in sophisticated virtual cockpits limits the simulation's utility and value to the entire aviation community. It is this conviction that leads to the central premise of this book:

As an aid to training, the parts of *Microsoft Flight Simulator* can be greater than the whole.

Put another way, since we in the flight training community agree that an aircraft cockpit is a less-than-ideal place to learn new concepts and skills, why do we so often insist on encumbering an inexpensive, multipurpose, portable, PC-based simulator with costly, complicated accessories and then use it only as if it were a complete cockpit?

the Flight Simulator paradox

Ironically, it is the powerful illusion of flight provided by the latest versions of *Microsoft Flight Simulator* that leads to difficulties the aviation community has in finding the best way to use PC-based simulations in flight training applications.

When Flight Simulator appeared in the early 1980s—the "green sticks era," with graphics that featured both horizontal *and* vertical lines—it was easy to dismiss the program as just a game.

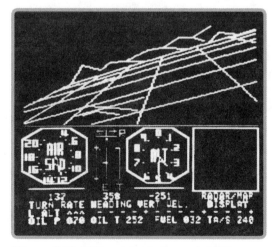

Microsoft Flight Simulator in the early 1980s

Today, however, *Microsoft Flight Simulator* replicates in stunning detail the experience of sitting in a cockpit. Its authentic instrument panels and visual display, even when limited to a straight-ahead orientation on a typical computer monitor, are more detailed and realistic than the cockpit environments and out-the-window views offered by many purpose-built and approved flight training devices, to say nothing of pioneering simulators like the Link Trainer and ATC 610. Combine the visuals in Flight Simulator with the worldwide database of airports and navaids; dramatic, real-time weather effects; interactive ATC; and other features, and it's hard to think of Flight Simulator as anything but a substitute for time at the controls of a real aircraft. In other words, if it looks like a cockpit, it ought to be used like a cockpit.

The latest versions of Microsoft Flight Simulator reproduce the aviation environment in stunning detail.

But new technologies like the DVD and the Web suggest more creative and flexible models of how to apply tools like *Microsoft Flight Simulator* to the challenges of flight training. These inventions, although developed for other purposes, have transformed the educational process and rescued students from interminable exegeses of the FARs and "death-by-PowerPoint" presentations.

I do not mean to suggest that DVD-based ground schools, GPS simulators, and Web-based courses are always better than classroom time or one-on-one sessions between students and instructors. Yet these new technologies offer advantages that until recently were difficult or very expensive for typical flight schools to deliver: flexibility, portability, convenience, time savings, lower costs, adaptability, standardization, and focus on specific tasks and procedures.

For example, students can watch an instructional DVD about a GPS unit whenever they have time and wherever a DVD player is available. They can replay the same "lecture" as many times as necessary to learn the material, skip sections they already know, jump to specific areas for review before a particular lesson, and so forth. A trip to the airport is not required, and the flight school does not have to provide equipment, space, or supervision to support such on-demand training. Instructors need not repeat the same basic information to dozens of students, draw complicated, static diagrams on white boards or teach to the lowest common denominator in a class. Instructors also don't have to worry about omitting important details because time is short, they lack recent experience with a particular unit, or lesson plans aren't complete. Instead, teachers can devote their valuable one-on-one or classroom time to emphasizing and clarifying specific information, adapting their teaching to each student's needs, supervising hands-on practice, and evaluating their students' knowledge.

The same advantages also apply to other high-tech tools, such as the interactive online courses offered by the AOPA Air Safety Foundation.

Many companies that provide training in sophisticated, certified flight training devices have made self-directed, interactive learning the foundation of their programs. They have replaced much traditional classroom time with self-paced, computer-based training sessions. Pilots don't begin sessions in full-motion simulators until they've become thoroughly familiar with cockpit layouts, procedures, and certain components like flight management computers. Usually they practice specific scenarios on functional mockups and in procedural trainers that lack visual displays and other accessories.

This interactive approach to teaching is increasingly important because many of today's students expect all learning to be interactive, engaging, and portable, and flight schools and instructors that offer such experiences have an edge in attracting and retraining customers.

Modern "flight training" often begins with interactive lessons that focus on specific components. (Alteon photo)

the mental game

The task-specific, interactive approach to training also addresses what my experience as instructor suggests is another major obstacle to effective training, especially in a typical general aviation program: mastering the concepts behind flying.

By "concepts" I do not mean theories like Bernoulli's Principle (how—or if—Bernoulli's equation applies to airfoils is a subject for another book). I use "concepts" to mean the principles behind such topics as navigation, operating the aircraft equipment, specific flying procedures, applying abstract knowledge of the weather, and so on.

In his inimitable way, hall-of-fame catcher Yogi Berra encapsulated this training predicament when he said of his sport, "Ninety percent of baseball is mental; the other half is physical." I've never tried to hit a major-league fastball, but having taught people to fly and struggled to learn the finer points of aviation myself, I think that Yogi's aphorism applies equally well to flying.

In my experience, many of the difficulties faced both by new students and experienced, but rusty, pilots stem from a lack of understanding of the principles behind the tasks they are trying to master. Obviously, many pilots could benefit from practicing more crosswind landings, stalls, and other fundamental tasks involving basic motor skills. It is also true that understanding the aerodynamics of side slips alone can't keep you out of the weeds when a gusty crosswind shoves you off centerline.

To master some skills, you must practice them. Muscles need training and regular exercise. Still, unless you're trying to win aerobatic competitions, the physical part of flying usually isn't that tough. Nonetheless, you can't fly standard traffic patterns, make precise instrument approaches, or impress the controllers with textbook holding-pattern entries unless you understand what you're supposed to do before you start wiggling the controls. This is where using Flight Simulator as more than a complete, virtual airplane cockpit comes in to play.

Quick question: You are flying direct to a VOR from the southeast to hold northwest on the 310 radial, right turns. What type of entry should you fly, and what's the heading to turn to after you cross the fix? Holding patterns are mostly a mental—not a flying—challenge.

obstacles to learning

Consider a typical instructional flight. The lesson begins with preflight briefing on specific tasks. On the ground, the student seems to absorb the abstract theory and the steps required to accomplish the tasks for the day. Yet because the only tools typically available to an instructor are a white board, pictures in a book, and perhaps a model, the student doesn't actually rehearse skills or see the concepts in action until after takeoff.

Furthermore, the student must accomplish many intervening tasks before practicing the specific goals of the day's lesson: dispatch the aircraft; complete a preflight inspection; run checklists; follow ATC instructions to taxi, take off, and exit the traffic pattern; navigate to the practice area; look out for other aircraft; avoid controlled airspace; and endure the instructor's incessant patter about altitude, heading, trim, right rudder, and power settings. When the student is finally ready to practice a maneuver that was discussed 30 minutes ago, intervening events and stresses too often obscure the fundamental concepts.

using flight simulator as a training aid

Similar obstacles often frustrate an instructor's efforts to identify and correct a student's mistakes. When instructor and student finally return to the briefing room for a post-lesson review, the student (and often the instructor) may no longer remember specifically what went wrong in the practice area because of interruptions to the lesson, events on the flight back to the airport or post-flight chores such as securing the airplane and tracking down the fuel truck.

Galvin Flying

A typical session in a flight training device often follows a similar script, omitting or accelerating some routine steps, but still emphasizing the simultaneous operation of the flight controls, avionics, and other systems as if in a real airplane. Even using the "freeze" button is frequently discouraged as "cheating."

More to the point, cranking up a flight training device just to demonstrate how to tune a radio or to let a student observe VOR needles in action is often frowned upon as a wasteful, cumbersome, expensive use of a valuable resource. Therefore, even at flight schools equipped with and savvy about using simulators, it's usually back to the white board for the theory and then directly into the airplane or simulator for hands-on practice. That process provides few intermediate opportunities for students to experience and rehearse discrete tasks, and to master each task before moving on to the next.

This approach too often filters down to those who have invested in accessories to create Flight Simulator pods. They utilize Flight Simulator only as if it were a full-fledged flight training device. Instructors insist that it be used as a cockpit and "flown" as an aircraft, not employed as an interactive teaching tool that can play many roles. As a result, we too often teach distinct concepts, such as tracking VOR radials, as part of the entire, complex task of flying. It is as if we were training novices to juggle by insisting, from the very first practice session, that they simultaneously keep six balls in the air.

Flight Simulator: a "swiss army knife"

The key to putting *Microsoft Flight Simulator* to its most effective use throughout the flight training process is to think of it as multi-purpose tool like a Swiss Army Knife, not just as a "flight simulator." A Swiss Army Knife is not necessarily the best knife, corkscrew, or can opener, but it's portable and less expensive than a comparable set of implements and gadgets. Because in many situations it's the only tool available, it can also be indispensable. When viewed with a little imagination, Flight Simulator has the same qualities.

For example, suppose you are working with a pre-solo student and it is time for the fledgling to take over using the radios. Ideally, you would have a working mockup of the avionics so that, in the comfort of a briefing room, the student could practice finding frequencies on a chart, twisting the tuning knobs, pushing the standby-active flip-flop switch, and selecting a transmitter on the audio panel.

A couple of decades ago, you could conduct this lesson in the cockpit, even with the airplane sitting cold in the chocks, because radios had mechanical frequency displays. Today's digital avionics must be turned on if you want to see the frequencies and use the controls, so you must hook up an external power supply, run down the battery, or watch the Hobbs meter click off dollars while you and a student sit in the cockpit with the prop turning. If your flight school has a well-equipped flight training device or simulator, you could start it up and practice using its avionics, but like airplanes, most FTDs have currency-counting meters, and they are in demand for "real" training.

Now, imagine using *Microsoft Flight Simulator* for that lesson. You sit down with your student at a PC or laptop running Flight Simulator and, with the appropriate charts and directories, you explain finding frequencies, tuning radios for different situations and how to use the audio panel. This lesson need not include any "flying"; you can accomplish all these goals with Flight Simulator parked on the ramp. Or you can use the autopilot to handle the flying chores so that you and your student can focus on the core task.

You can use the avionics stack in Flight Simulator to rehearse tuning and using radios.

When using Flight Simulator for an exercise like this, you remove a lot of pressure. You are not sitting on the ramp burning gas and blocking a taxiway. No Hobbs meter turns over. The student isn't distracted by a stream of rapid-fire communications to and from other aircraft. You can carefully demonstrate procedures and then observe the student, taking as much time as necessary to repeat scenarios and integrate tasks with cockpit flows and checklists. Throughout the lesson, you and the student can verbalize typical exchanges with ATC, make announcements on a virtual CTAF, and practice calling Flight Watch.

Is Flight Simulator the perfect tool for this lesson? Of course not. Using the mouse to "rotate" tuning knobs and "push" flip-flop buttons does not precisely duplicate using your fingers to perform those actions. The radios in Flight Simulator may not look exactly like the boxes in the airplanes you fly. Even so, the avionics in Flight Simulator generally represent the way most current radios look and function, and they can serve as an interactive replacement for a radio stack anywhere and anytime students have access to a computer and Flight Simulator.

Even with a basic setup, Flight Simulator can be an effective teaching tool.

Here is another way to think of this approach to using Flight Simulator. Not every car door has the same type of handle, but differences in hardware don't keep people from applying their door-opening knowledge from entering a variety of different cars. The reason is that we know pulling or pushing some piece of metal in the upper-right hand corner of a door opens it. The same principle applies to using a mouse with Flight Simulator. Move-and-click some area of the radio, and a frequency changes. There is enough similarity to transfer the basic idea of "changing frequencies" to a real radio.

interactive chair flying

Using Flight Simulator to rehearse using radios and for developing other skills, such as interpreting cockpit instruments and applying abstract knowledge about navigation, is really an updated example of "chair flying," or "visualization". This technique has been useful in flight training since Wilbur and Orville first took to the air. Including Flight Simulator in your training toolbox just makes mental rehearsal more vivid and effective.

To borrow a definition from "Principles of Learning" in the FAA's *Aviation Instructor's Handbook*, the interactivity that Flight Simulator brings even to a non-flying lesson is an application of the principle of "intensity": "A vivid, dramatic, or exciting learning experience teaches more than a routine or boring experience." [1]

Using Flight Simulator as a multi-purpose teaching aid also helps instructors overcome another common obstacle that inhibits effective training, as noted in the same section of the *Aviation Instructor's Handbook*:

> *In contrast to flight instruction and shop instruction, the classroom imposes limitations on the amount of realism that can be brought into teaching. The aviation instructor should use imagination in approaching reality as closely as possible.*

As we will see in Chapter 3, "Best Practices for Using Flight Simulator," employing Flight Simulator creatively outside the confines of the cockpit can overcome many "limitations on realism" in ground instruction, and bring real-time interactivity to almost any lesson in a training syllabus.

1 *Aviation Instructor's Handbook* (FAA-H-8083-9), 1–5

concerns about using pc-based simulations

As I noted at the beginning of this chapter, many instructors, students, and pilots recognized early on that PC-based simulation could complement flight training and help all aviators maintain proficiency. Just as quickly, however, experts in the aviation community raised concerns about how PC-based simulators might adversely affect training and safety. It is important to understand and address these issues before instructors, students, and pilots employ any device in training and proficiency programs.

negative transfer

Many concerns about using PC-based flight simulators in flight training fall under the broad heading of "negative transfer." The dictionary defines that phrase as, "The interference of previous learning in the process of learning something new."[2] Negative transfer is related to the important principle of "primacy," which, according to the *Aviation Instructor's Handbook*, asserts that, "[W]hat is taught must be right the first time. For the student, it means that learning must be right…The first experience should be positive, functional, and lay the foundation for all that is to follow."[3]

These doctrines often reinforce instructors' insistence on using Flight Simulator only with a yoke, rudder pedals, power levers, and other controls. This is a well-intentioned, informed inclination. It makes sense if the primary goal of using Flight Simulator is to teach stick-and-rudder skills while practicing a complete set of flying tasks exactly as they should be performed in the cockpit.

It is hardly surprising that people who develop their flying skills using Flight Simulator without proper instruction typically form several bad habits. When placed at the controls of a real airplane, they quickly demonstrate gaps in their knowledge, understanding, and performance of basic tasks. They often must "unlearn" incorrect control inputs, misunderstandings about systems and basic operating procedures.

I have given introductory flights to many Flight Simulator enthusiasts who have had no previous flying experience. They are comfortable in the cockpit (even if there's less space for snacks and the other creature comforts that they enjoy in their home cockpits), and they are generally familiar with the instruments, controls, and basic avionics. These virtual aviators also often understand the rudiments of airport operations, air traffic control, and the aviation environment. Once in the air, however, it is immediately apparent that they focus too much on the instruments and must be reminded to look outside. Usually they lack basic stick-and-rudder skills, and are rapidly overloaded by the demands of juggling airplane control, collision avoidance and communication in a real airplane moving in three dimensions. At the end of that first flight, these new pilots often confess that virtual flying certainly helped, but did not fully prepare them for flying an actual airplane.

Nevertheless, fledgling virtual aviators are also far ahead of those with no prior aviation experience. As many testimonials from students and instructors have noted, with proper guidance and supervision, virtual aviators usually adjust rapidly to real-world flying and make good progress in their training.

2 *The American Heritage® Dictionary of the English Language,* 4th ed. Boston: Houghton Mifflin, 2000
3 *Aviation Instructor's Handbook* (FAA-H-8083-9), 1–5

The keys to minimizing negative transfer when using Flight Simulator as a training aid apply to all phases of training. They include:

- Clearly identifying which skills and tasks Flight Simulator is best suited to develop, especially during the early phases of flight training.
- Focusing on Flight Simulator's strengths and using it to learn and master specific, discrete skills. For example, regardless of the type of controls attached to a PC, Flight Simulator cannot duplicate the precise yoke or stick inputs (i.e., actual forces and amount of movement) required to control an airplane. However, Flight Simulator can help an instructor explain and demonstrate the general purpose and effect of the controls. It is particularly useful for specific procedures such as the proper use of elevator trim. The differences between the way the aircraft in Flight Simulator and real airplanes respond to the controls become less important as pilots develop a feel for flying.
- Modifying the tasks rehearsed in Flight Simulator as a student gains experience. For example, with a private pilot student, use Flight Simulator to illustrate and practice abstract tasks such as VOR navigation instead of basic flying skills. On the other hand, an IFR student who has become proficient at basic flying skills and who has used Flight Simulator and its autopilot to learn the fundamentals of a specific IFR procedure may be ready "fly" an entire approach to prepare for a session in a full FTD or an airplane.
- Understanding and clearly explaining the differences between the way specific tasks are accomplished in Flight Simulator and in a real aircraft. Some disparities between Flight Simulator and a real cockpit are trivial—for example, displaying the avionics in a pop-up window. Other distinctions are more important—for example, proper engine-starting procedures.
- Recognizing that no single approach to using Flight Simulator will work for all pilots at every stage of training. Chapter 3, "Best Practices for Using Flight Simulator" discusses many specific ideas for using Flight Simulator in a typical training program.

the learning environment and virtual aviators

Virtual aviators who make a concerted effort to study and practice the lessons included with Flight Simulator can learn a great deal about the principles of flight. The problem is that upon completion, they don't know how much they actually understand about flying. In particular, they often don't have the cockpit "context" to organize the material they've learned in the lessons. In other words, they learn bits and pieces in Flight Simulator, but once in an airplane they may not have immediate mental access to this information.

When introducing virtual flyers to their first real airplane flight it's often best to help them connect with what they know by asking such questions as, "OK, what will happen when we apply power for takeoff? What will the airspeed do? What should the airspeed read when we decide to rotate the airplane? Using your hand, show me the angle that you will pitch to in order to establish the proper climb attitude." And so on. This approach helps virtual aviators connect with what they already know.

In other words, the knowledge that experienced virtual aviators acquire is subject to what psychologists call State Dependent Learning. Virtual aviators learned about flying fundamentals in a "simulated" state (a calm, relaxed, home environment). When placed in the cockpit of a real airplane, however, they try to apply that knowledge while bombarded by a new set of potentially overwhelming stimuli. The concept of State Dependent Learning argues that it's difficult to access information learned in one environment and level of stress when you're suddenly placed in a different, much more dynamic, stimulating situation.

realism: flight models and controls

Many pilots and instructors dismiss PC-based simulations because the virtual aircraft in Flight Simulator do not "fly" like real airplanes. Even if these aviators continue using Flight Simulator, their frustrations with simulated flying often interfere with any benefits they might gain from the experience. Much of this disappointment with virtual flying stems from confusion between the "flight model" and the subjective "feel" of how a simulated aircraft responds to the controls. The issue is compounded by a lack of sensory input and the limited view typical of PC-based flying.

It is important to understand the difference between the "flight model" and the "feel" of flying that results from moving the controls. The simulation engine in Flight Simulator uses a well-established six-degrees-of-freedom (6-DOF) flight model (roll, pitch, and yaw rotations plus lateral, longitudinal and vertical translations) and standard equations of motion to define the aircraft's movement in space. A set of classic aerodynamic coefficients combined with information about an aircraft's mass and geometry, plus data from a set of tables, produces the forces and moments acting on the aircraft. The simulation engine shows the effects of those forces on the cockpit instruments and updates the out-the-window view. (Technically inclined readers can find a detailed description of the flight model in the article "Aircraft Simulation Techniques Used In Low-Cost, Commercial Software," available on the official *Microsoft Flight Simulator* website.)

The flight models in *Microsoft Flight Simulator* reproduce the performance and general characteristics of specific aircraft quite well. That is, if you establish a pitch attitude and set engine power and aircraft configuration (flap setting, landing gear position, weight, and so forth), the airspeed and rate of climb or descent accurately reflect the performance of the real aircraft usually to within a few percent. Of course, the aircraft in Flight Simulator are not subject to many variables that affect real airplanes. Virtual airframes and flight controls do not sag out of rig; bugs, dirt, and dings don't add drag or affect handling; and simulated engines always produce rated power. Anyone who regularly flies aircraft in a flight school fleet knows that minor variations in performance and handling are common, even among airplanes of the same make, model, and year of manufacture.

The way a simulated aircraft "feels," however, depends on many factors beyond the accuracy of the flight model. As noted in Chapter 1, "About this Book," the yokes and joysticks typically available lack the fidelity to reproduce perfect control input and response, especially across a range of aircraft types. In addition, there is tremendous variation in the responsiveness and sensitivity of the joysticks and yokes produced by different manufacturers.

For example, a popular flight yoke that I often use for testing has a total fore-aft range of approximately 3 inches. It rotates some 45 degrees to either side on the "aileron" axis. The fore-aft range of movement of the yoke in a late-model Cessna 172, however, is about 6 inches, and the yoke turns some 60 degrees to either side. The elevator control in another aircraft simulated in Flight Simulator, a Beechcraft BE58 Baron, also has a wide range of motion compared to a typical PC controller, and it includes a down spring that adds a heavy forward force on the yoke.

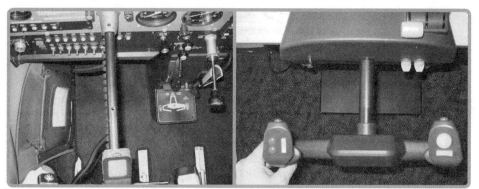

PC controls don't reproduce the full range of motion of real aircraft controls.

Now consider that Flight Simulator must also adjust inputs from a given yoke or joystick to work across a wide range of aircraft, from light prop aircraft to heavy jet transports. One quarter inch of aft-yoke movement when flying a virtual Cessna 172 applies a specific angular deflection of the elevator. Depending upon the aircraft's true airspeed, center of gravity, elevator trim setting, and the force of the air being blown over the horizontal stabilizer by the propeller, an input results in a simulated pitch force and movement of the aircraft's nose up or down. That same one-quarter-inch movement must also deliver the correct response when flying, for example, a virtual Boeing 747.

The control loads that a pilot feels when flying a real airplane are similarly difficult to reproduce with PC controls. Most yokes and joysticks incorporate springs or other mechanisms to create resistance and control "feel." More expensive "force-feedback" yokes and joysticks use motors to simulate variations in control loads as airspeed, power, trim, stabilizer setting, and other factors change. However, because of cost and other factors, these devices produce feeble effects compared with the forces a real pilot encounters. This is especially true during maneuvers such as a go-around with full flaps from an idle-power approach with the aircraft C.G. at its aft limit.

The absence of kinesthetic inputs and the limited field of view associated with flying at a desk impose additional constraints on the subjective "feel" of a PC-based flight simulation. As they gain experience, pilots develop subconscious reactions to the movements of an aircraft in flight. They feel the airplane "sag" as speed decays during a landing flare. A turn pushes them into the seat. Increasing or reducing power subtly presses them back or tugs them forward. They use these physical cues to make constant adjustments as they fly and to gauge how much control force is needed for a maneuver. In addition, as anyone who has watched an IMAX® film can attest, our brains are so attuned to movement in the real world that wide-angle, vivid visual cues can generate a false sense of motion even when we're sitting still. The large, wrap-around displays in "real" flight simulators similarly induce powerful illusions of movement as the pilot manipulates the controls, even when the motion systems are turned off.

The absence of all these sensory inputs while desk-flying contributes to the disconnect between control inputs and aircraft reaction that many people experience when using Flight Simulator. In fact, flying Flight Simulator with a typical display (oriented to a single, straight-ahead view) is a lot like flying at night, when depth perception, peripheral vision, and other visual cues are limited or unavailable.

These considerations suggest that Flight Simulator is not the best tool for *developing* basic stick-and-rudder skills, especially for students or pilots with limited experience flying real aircraft. In fact, I usually keep such students away from the PC controls until they gain experience with how control inputs relate to aircraft movement and the sensations of real flight. Later, we can "fly" Flight Simulator with a better understanding of how it relates to real aircraft.

However, Flight Simulator can still be useful with new students even when the lesson includes details about the flight controls. For example, instead of holding an aircraft model to illustrate how the primary controls rotate an airplane about its axes, I can use realistic joystick or yoke inputs to bank the wings and pitch the nose up and down. The student can see important principles in action, both from outside the airplane (in spot plane view) and from the pilot's seat. I can use Flight Simulator to demonstrate that by maintaining specific control input, the corresponding roll or pitch movement continues. Likewise, I can show more subtle effects, such as how changing power or extending or retracting flaps induces pitch changes.

The certification standards for flight training devices established by the FAA also address the relationship between the fidelity with which a flight model reproduces the characteristics of a specific aircraft and a device's value as a training tool. The FAA certifies flight training devices and simulators in categories or levels, based on the complexity and realism of the device. For example, the FAA currently recognizes seven levels of flight training devices, and it certifies simulators as level A, B, C, or D devices. Each of these certification levels defines specific requirements for the fidelity of both the cockpit environment and the flight model that drives the simulation.

level	cockpit	aerodynamic model	control loading	sounds system	motion system	visual system
general requirements for flight training devices by levels (from AC 120-45A)						
1	reserved	reserved	reserved	reserved	reserved	reserved
2	generic (open or closed)	generic	no		optional	optional
3	generic (closed)	generic	yes	yes	optional	optional
4	specific for make/model (open or closed)	not required	not required		optional	optional
5	specific for make/model (open or closed)	generic	no		optional	optional
6	specific for make/model (open or closed)	specific for make/model	yes	yes	optional	optional
7	specific for make/model (open or closed)	specific for make/model	yes	yes	optional	optional

"Simulators, Logging Time, and FAA Approval," later in this chapter, has more to say about approved training devices, but with respect to flight models, it's worth noting here that, according to FAA Advisory Circular 120-45A, "Airplane Flight Training Device Qualification":

Level 2, 3, and 5 flight training devices do not require a specific aerodynamic model;…In the absence of a specific model, these devices may use a generic model typical of [a] set of airplanes…For example…a light twin or single engine airplane flight training device must demonstrate performance typical of the respective set of airplanes. The aerodynamic model may be one representing an actual airplane within that set of airplanes or it may be created or derived using the same mathematical expressions as those used in a specific model, but with coefficient values which are not obtained from flight test results for a particular airplane. Instead, the coefficient values could be fictitious, but be typical of the set of airplanes replicated. The reference validation data could then be created by doing a computer simulation using these fictitious coefficients. A generic model may also be acquired from public domain resources or it may be a composite of various models, none of which is complete within itself.

The fidelity of their flight models notwithstanding, the Level 2, 3, and 5 flight training devices described in the table above (a Level 4 device does not even require a flight model) are all approved by the FAA for ranges of tasks related to the training, proficiency and certification of pilots. The FAA and the aviation industry recognize that the value of these devices lies in their ability to "permit learning, development, and practice of skills and cockpit procedures," not necessarily in honing a fine touch on the yoke, rudder pedals, and throttle.

stalls, spins, and other maneuvers

Debates about the realism of Flight Simulator and its flight models often focus on stalls, spins and related maneuvers that involve the corners of an aircraft's flight envelope. As explained in the previous discussion, many factors limit the utility of a desktop simulator as a tool for developing stick-and-rudder flying skills, especially outside the range of basic flight maneuvers.

The inherent difficulty of developing an aerodynamic model that accurately replicates stalls and spins complicates the problem of simulating such flight maneuvers. A history of the NASA Langley Research Center states the problem succinctly:

> The aerodynamic characteristics of most aircraft configurations become extremely nonlinear and ill behaved at angles of attack beyond stall. Thus, the prediction and analysis of stall/spin behavior have not been amenable to theoretical methods. [4]

NASA continues to use a special vertical spin tunnel to test scale models of aircraft. They also rely on flight tests to validate the data from those experiments, because, according to the same report, "The final, definitive answers in any analysis of stall and spin behavior are provided only by flight tests of the full-scale aircraft."

Furthermore, the Normal category aircraft that comprise the majority of the personal and flight-training fleets do not undergo extensive spin tests, and therefore little data is available to support detailed computer modeling of such aircraft in the stall/spin corners of the flight envelope. As noted in the *Airplane Flying Handbook* (FAA-H-8083-3A):

> …Normal category airplane certification only requires [that] the airplane recover from a one-turn spin in not more than one additional turn or 3 seconds…[and] 360° of rotation (one-turn spin) does not provide a stabilized spin…There are no requirements for investigation of controllability *in a true spinning condition for the Normal category airplanes. The one-turn "margin of safety" is essentially a check of the airplane's controllability in a delayed recovery from a* stall.[5]

The most important issue related to the use of Flight Simulator in the stall/spin arena, however, isn't the realism of or limitations associated with a flight model. Stalls—and especially spins—involve many unusual, dynamic physical sensations and visual cues that make rehearsing the maneuvers while sitting at a desk at 1G of dubious training value. The previously discussed limitations of PC controls exacerbate the problem of "flying" Flight Simulator through such maneuvers.

That said, instructors can use Flight Simulator as if it were an interactive model or other teaching aid to effectively demonstrate some elements of stalls and spins, such as the procedure for entering and recovering from basic varieties of stalls.

4 Chambers, Joseph R., *Concept to Reality: Contributions of the NASA Langley Research Center to U.S. Civil Aircraft of the 1990s.* NASA SP-2003-4529. October 2003. "Spin Technology" http://oea.larc.nasa.gov/PAIS/Concept2Reality/spin_technology.html
5 *Airplane Flying Handbook* (FAA-H-8083-3A). 4-15.

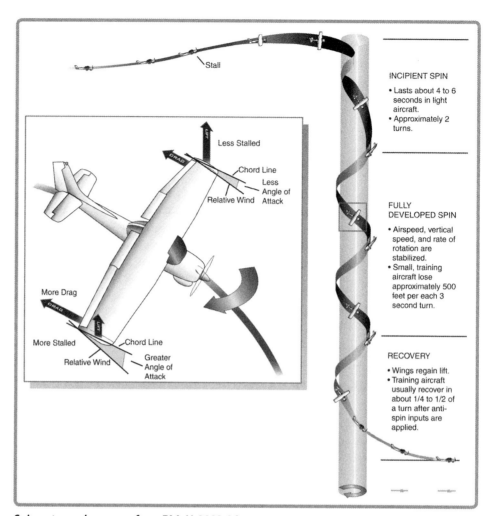

Stall

INCIPIENT SPIN
- Lasts about 4 to 6 seconds in light aircraft.
- Approximately 2 turns.

FULLY DEVELOPED SPIN
- Airspeed, vertical speed, and rate of rotation are stabilized.
- Small, training aircraft lose approximately 500 feet per each 3 second turn.

RECOVERY
- Wings regain lift.
- Training aircraft usually recover in about 1/4 to 1/2 of a turn after anti-spin inputs are applied.

Spin entry and recovery from FAA-H-8083-3A

Dealing with the "realism" issue, then, is largely a matter of understanding what tasks Flight Simulator is best suited for as individual students and pilots move through each stage of their training. Flight instructors and pilots must apply their judgment and experience to determine how and when to "fly" aircraft in Flight Simulator. They must ensure they understand the value and limitations of the virtual "stick time" they accumulate.

flying "my airplane"

Many pilots understandably want to fly a specific airplane during their sessions with Flight Simulator. They count on practicing in a type that matches the performance and handling characteristics of the aircraft they fly, and they naturally see value in using an instrument panel and avionics stack that closely resemble the layout and functions they are accustomed to. Unfortunately, however, if an airplane and a specific instrument panel/avionics configuration isn't represented by one of the standard aircraft, pilots too often dismiss Flight Simulator.

The fleet of aircraft in Flight Simulator developed primarily to provide a variety of experiences for virtual aviators. Over the years, the collection has grown from a few basic models to an array that represents a variety of categories and classes of aircraft, from a Piper Cub to a Boeing 747. The mix includes single-engine and twin-engine piston aircraft; a turbocharged, high-performance, single-engine airplane; airplanes equipped with tailwheels and floats; and jets. Virtual aviators can also purchase or download thousands of add-on aircraft created by enthusiasts from around the world. As with an elaborate cockpit, however, a particular aircraft make and model isn't essential for using Flight Simulator effectively as a training aid.

First, note that the standard aircraft in Flight Simulator represent broad classes of aircraft that share similar performance and handling characteristics. A Cessna 172, although it has a high wing, operates with almost the same performance characteristics as a low-wing Piper Warrior. Of course, those aircraft have important differences, such as the operation of the fuel system and the sight-picture from the pilot's seat. They also react differently to some actions, such as flap extension. Additionally, the engine and systems-monitoring instruments, switches, and other controls are in different locations.

Nevertheless, a generic aircraft can work just as well as an exact duplicate of a specific make and model. Consider the training that goes toward developing an effective instrument scan, learning about VOR navigation or practicing the elements of a VFR cross-country flight. Indeed, there are dozens if not hundreds of isolated skills that can be addressed with Flight Simulator.

Modern aircraft instrument panels use the same basic layout.

In fact, the instrument panel layout that the general aviation industry converged on and adopted as a standard in the 1970s was developed because it supported a more efficient, useful scan and helped pilots transition easily between aircraft. The design and operation of basic avionics has followed similar convention. Though produced by different manufacturers, communication and navigation radios and VOR displays look and work very much alike (technologically advanced "glass cockpit" aircraft notwithstanding).

Professional and military training programs have long recognized that valuable flight training can be accomplished by the use of proxies. For example, the U.S. Air Force uses a military version of the Beechcraft 400 business jet to help prospective transport and tanker crews develop and maintain proficiency in such fundamental skills as cockpit coordination and IFR procedures. The differences in performance, maneuverability, and handling between the T-1A Jayhawk and the C-17 Globemaster III, C-5 Galaxy, KC-135 Stratotanker, and other aircraft that pilots eventually will fly aren't important during the early phases of pilot training.

T-1A Jayhawk (U.S. Air Force photo by Terry Wasson)

C-17 Globemaster III *(U.S. Air Force photo by Kevin J. Tosh)*

Similarly, NASA astronauts zoom around in T-38 trainers to help them stay sharp (and, of course, look cool) between missions. In part, this is because time in the Shuttle Training Aircraft (Gulfstream business jet modified to simulate the orbiter's landing approach) is expensive and reserved for practicing procedures related to that single phase of flight.

These programs recognize that time spent in *any* cockpit has value because it keeps pilots mentally sharp and reinforces procedures and basic flying skills. Properly focused practice in an aircraft—even one that does not exactly duplicate the handling characteristics, performance, and cockpit layout of a pilot's primary aircraft—is better than limiting training to rare opportunities when a specific aircraft may be available.

The same considerations apply to many flight training devices. As noted in Chapter 1, "About this Book," I often fly an expensive, FAA-approved simulator at the flight school where I teach. The simulator accurately reproduces the *performance* of each of the aircraft it emulates (a C172 Skyhawk, BE-36 Bonanza, and PA-44 Seminole), but it doesn't *feel* much like any real airplane. Its generic light-airplane cockpit layout only generally resembles the arrangement of the instruments and controls of the three aircraft that it simulates.

Frasca 142 FTD cockpit

Cessna 172S cockpit

More support for the benefits of general training comes from accident statistics, which consistently illustrate that many common accidents and safety issues are not strongly correlated with specific types of aircraft. Pilots often get into trouble for the same reasons. They fly into bad weather, crash into terrain and run low on fuel after becoming disoriented. In some cases they just have poor or rusty fundamental skills. When properly included in training and recurrency programs, Flight Simulator can help address such issues.

Obviously, using an exact duplicate of a specific aircraft and its systems for all training related to that aircraft would be ideal. But Flight Simulator, like that Swiss Army Knife, can often prove itself a versatile—even invaluable—tool, especially when compared with the option of not using of simulation at all.

self instruction

Every instructor has faced the problem of helping a student break bad habits or adjust to a new way of doing things. Often such issues stem from the fact that most flight schools and instructors adopt variations on the common procedures that all pilots must learn. Pilots also develop their own flying styles as they gain experience. Because FAA regulations, insurance provisions, and FBO rules mandate that pilots receive initial training and get periodic checkups with instructors, there are many opportunities during a typical pilot's flying career to uncover and correct bad habits. This is not to deny that some certificated pilots show up for checkouts, flight reviews, or advanced instruction with serious misconceptions and unconventional flying techniques. They presumably acquired these habits sometime during their flight training. Few pilots (outside the remoter parts of Alaska) are truly self-taught and left to develop their skills only through their personal experiences.

Virtual aviators, however, are another matter. They may cultivate bad habits and harbor misconceptions because they usually have developed their flying technique without the guidance of an instructor or get conflicting, incomplete advice from other enthusiasts. Often virtual aviators are tempted to launch Flight Simulator before they understand basic concepts, and they compound bad habits through repetition without opportunities for correction.

Including Flight Simulator in a flight training program similarly opens opportunities for loosely supervised students to veer off the proper learning course. They may acquire bad habits, reinforce misunderstandings, and impede their progress by jumping ahead of a carefully designed syllabus.

The solution here is also a familiar one. Instructors and pilots should use Flight Simulator throughout a training or proficiency program. Instructors should encourage students to include Flight Simulator in home-study sessions. Yet all virtual aviators should use Flight Simulator wisely, with a structured plan. There should be provisions for careful guidance, and a full understanding of how best to apply the tool at each stage of the learning process. Certificated pilots who complement their normal flying with Flight Simulator sessions should also ensure that they discuss questions and flying techniques with their instructors so that issues can be identified and resolved before they become serious problems.

instrument panel fixation

Virtual aviators who start their flying careers on Flight Simulator tend to fixate on the instrument panel, for all the reasons explored earlier in this chapter. Of course, this problem isn't just limited to pilots who "fly at their desks." Many new pilots stare at the instrument panel because they have difficulty relating the view out the window to the controls in their hands, so they must expend extra mental cycles to figure out which dials and gauges are which and how to interpret them. This habit appears in almost all new pilots except for those who learn in airplanes equipped with only the minimum required instruments (and even they are quickly seduced when they move up to aircraft with a full complement of gauges or a "glass cockpit").

Like most flight instructors, I have observed this problem for years with students and pilots who have little or no exposure to simulation of any kind. During instruction, flight reviews, and aircraft checkouts, I frequently have to remind the person in the left seat to look around and clear the area before and during routine turns, to say nothing of stalls, lazy eights, and other maneuvers. The problem is so common that it has long been featured in training texts such as the FAA's *Airplane Flying Handbook*, which often identifies common errors such as "attempting to execute the turn solely by instrument reference." Collision avoidance—looking outside—also prominently appears in the "Special Emphasis Areas" of the introduction to the FAA's Practical Test Standards. Therefore, panel fixation is not a new problem spawned only by PC-based flight simulators.

Flight instructors have long applied various methods to fix panel fixation, and some of these solutions work for virtual aviators as well, both while "flying" Flight Simulator and during time at the controls in the air.

For example, when taking a virtual aviator on a first flight, an instructor might say, "OK, you've learned to fly by referencing the instruments, but this flight will be different. Instead of looking at the AI, we're going to use the real horizon as the AI. Use all that you've learned about referencing the AI in the airplane, but this time, apply that knowledge to the real horizon."

A typical virtual pilot may know that, assuming constant power, raising the nose lowers the airspeed and lowering the nose increases the airspeed. This knowledge can be put to immediate use by asking the student, "Since you pitch up to climb at 80 knots, show me, by using your hand, the angle at which you think the nose has to be inclined to achieve that speed."

A student who has been properly and thoroughly exposed to the instruments (what and where they are, what they do, and how to interpret them) will spend less time head-down, puzzling over them in the air. It has always been a good idea, therefore, to review a cockpit poster or to sit in the cockpit with the airplane secure in the chocks, and while making engine noises, engage a student in a thorough tour of the gauges, dials, and controls.

Ironically, using Flight Simulator as part of such exercises can help prevent and ameliorate panel fixation. For example, an interactive tour of the flight instruments featured in Flight Simulator makes cockpit orientation more vivid and effective. This way, the instrument panel is shown in action and the student and instructor can discuss individual instruments and groups of gauges thoroughly, without distraction.

As discussed in Chapter 3, "Best Practices for Using Flight Simulator," you can hide all or part of the instrument panel and "cover" individual instruments to encourage students to shift their focus outside the cockpit. The Practice Flights described later in this book can also help instructors apply the principle of "integrated flight instruction" to show how the instruments on the panel relate to the big picture seen through the windows.

An instructor can even reinforce good collision-avoidance habits while using Flight Simulator with students. Demonstrate how to move the "hat" switch on the joystick or yoke to change the view from straight-ahead to look left and right. Then insist that students consciously "look" each time before they turn (even when the autopilot is on) while "flying" in visual conditions. This technique reminds me of stories my father told about his primary training in the Air Force. When walking the grounds near their barracks and classrooms, cadets had to call out, "Clear left!" or "Clear right!" before changing course on the sidewalks.

By the way, I have frequently observed the flip side of panel fixation when pilots sit down in front of Flight Simulator. They often stretch their necks and try to peer "over" the instrument panel on the computer monitor to get a better view of the outside world, especially during takeoff and landing. The same issue also appears when I put pilots in the front seat of my Extra 300L to begin a stall/spin lesson. The first question is inevitably, "Where's the ball?" (the inclinometer that helps a pilot maintain coordinated flight). The answer: "There isn't one up front." Students have to use the "gauge" they're sitting on to detect yaw, slips, and skids. Instrument dependency isn't limited to virtual aviators.

mouse and keyboard vs. real controls

Many pilots and instructors are concerned about negative transfer, awkwardness, and disruption of the "flying" experience inherent in using a keyboard and mouse instead of real knobs and switches to operate avionics and aircraft systems in a PC-based simulator.

As noted in Chapter 1, "About this Book," if a computer is equipped with just a basic joystick and a mouse, a keyboard is not required during a typical flight in Flight Simulator.

An instructor can alleviate concern about using a mouse in conjunction with a joystick and yoke to operate cockpit controls by explaining and demonstrating this key concept:

While using Flight Simulator, think of the mouse as an extension of your hand. If, in an airplane, you would reach for and push, pull, flip, or turn a control such as the throttle, flap lever, radio tuning knob or light switch, you can accomplish the same task by pointing to the control with the mouse and then clicking a mouse button or slowly rolling the mouse wheel forward or back.

You can use the mouse just like your hand to operate cockpit controls in Flight Simulator.

For example, to increase or decrease the power setting, move the mouse to put pointer symbol on the throttle. The pointer symbol will change from an arrow to a hand. Slowly and smoothly roll the mouse wheel forward to push the throttle in; roll the wheel back to pull out the throttle.

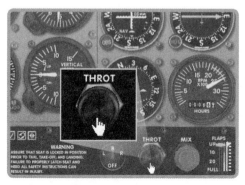

After you use the mouse to point to a control or switch, you can roll the mouse wheel forward and back.

Flight Simulator X re-introduces the "mouse as yoke" feature from early versions of *Microsoft Flight Simulator*. When you select this option, you can use the mouse as if it were a joystick or yoke. Flying with the "mouse as yoke" feature may interfere with using the mouse to operate some cockpit controls as described earlier in this section.

You can find more information about using the mouse in Chapter 4, "Flight Simulator Essentials."

Most students readily transfer this general concept to the cockpit. Using real controls is often then much easier than trying to maneuver the mouse pointer onto a small target area. Flight Simulator enthusiasts have no more trouble figuring out how to operate the throttle, flaps, switches, and other real cockpit controls than any other new pilot. An experienced instructor who conducted research with students who used PC-based simulators before taking to the air in real airplanes once reported, "No one ever climbed into the cockpit, looked around, and asked, 'Where's the mouse?'" (The appearance of sophisticated glass cockpits may soon make that story seem less amusing.)

Again, the limitations imposed by a computer-based interface need not become more of an obstacle to effective training than the differences between the controls in many approved simulators and the aircraft they represent. Rejecting PC-based flight simulation because it does not exactly replicate the controls returns students to the situation where learning takes place in the cockpit with heads-down, as they fumble with unfamiliar controls trying to catch up with the airplane.

Many effective training devices use generic controls.

the Flight Simulator database and IFR procedures

Over the years, many detail-oriented pilots have complained about the static database supplied in each version of Flight Simulator. In real-world flying, VFR charts are typically updated semiannually or annually, and IFR-approved GPS systems must be updated every 28 days. During the two or three years between versions of Flight Simulator, airports, runways, and navaids come and go. New instrument approaches are published and old procedures disappear. Obstacles sprout up and are torn down. The FAA adds and reconfigures (but rarely eliminates) controlled airspace. NOTAMs temporarily modify the minimums associated with specific instrument procedures.

Given that Flight Simulator is inexpensive and produced for a broad global audience, it isn't practical or economically feasible to supply regular, worldwide updates as published by aviation authorities. Companies like Jeppesen, which do provide such data refreshes, charge hundreds of dollars annually for their services. (To learn more about how many data sources are combined to generate the virtual world in Flight Simulator, see the articles on the subject on the official *Microsoft Flight Simulator* website.)

Changes in the real world may also frustrate enthusiasts who want Flight Simulator to reflect every detail during their virtual flights. Yet for several reasons, amendments to aeronautical data do not substantially affect Flight Simulator's effectiveness as a training aid.

First, understand that the Practice Flights associated with this book (and those that you create on your own) demonstrate and allow you to learn about and practice representative types of procedures: straight-in approaches, circle-to-land approaches, IAPs that include step-down fixes, transitions, course reversals, missed approach procedures, and other elements of instrument flying. The worldwide database of navaids and airports similarly supports practicing and honing a range of VFR skills, even if your home airport has acquired a new runway that's not in the Flight Simulator database.

Second, recognize that the fundamentals of flying an ILS or other type of approach remain the same, even if a specific, recently updated procedure acquires a new zig or zag. Likewise, the basics of en route navigation remain the same and can be learned and practiced anywhere, even as specific VORs and NDBs acquire new names or disappear.

Flight Simulator does not "include" instrument procedures, except for GPS approaches, which are based on waypoints in the database. Most instrument procedures technically exist only as descriptions on a chart (the official legal descriptions are embedded in the FARs), and you can fly them in Flight Simulator if you can receive the navaids and follow the published instructions. If a localizer/glideslope, VOR, or NDB frequency is in the Flight Simulator database, you can fly procedures based on those ground-based facilities.

Many virtual aviators have noted that the Flight Simulator database does not include SIDs and STARs. Because the database does not define these procedures, the ATC feature cannot include them in IFR clearances, and you cannot fly them when the autopilot is driven by the GPS. However, for training purposes, you can still fly procedures defined by ground-based navaids. These limitations of the Flight Simulator database are among the reasons that the Practice Flights accompanying this book do not use the ATC features in Flight Simulator.

To practice non-precision procedures such as VOR or NDB approaches, you can fly the latest courses and altitudes, even if they have changed from those in effect when Flight Simulator was released. An approach based on a localizer (ILS, LOC, or LDA procedure) has a defined course specified in the Flight Simulator airport and navaid database, so you must follow the courses established in Flight Simulator. But if other variables, such as minimum altitudes, change, you can use the latest charted values. In all cases, you must use the navaid frequencies included in the Flight Simulator database to receive and use signals from navaids.

The only significant restriction to the application of Flight Simulator in IFR training is the rapid publication of GPS-based IFR approaches. Unlike ground-based IAPs, GPS approaches (i.e., all the waypoints associated with an approach) must be in the database. Procedures published during or after the development of Flight Simulator are therefore not available. Nonetheless, the Flight Simulator database still includes a representative sample of GPS approaches (except for new WAAS-enabled procedures that include vertical guidance). You can use these procedures for practice around the world.

the Flight Simulator learning curve

Microsoft Flight Simulator runs on a PC, and that fact alone is enough to intimidate many pilots and instructors who think that they must learn yet another complicated tool before they can begin to enjoy benefits from virtual flying. Those who are familiar with Flight Simulator often think that setting up and configuring the simulation is not worth the trouble unless they plan to embark on a virtual cross-country flight.

Chapter 3, "Best Practices for Using Flight Simulator" and Chapter 4, "Flight Simulator Essentials" show that including Flight Simulator even in brief lessons and practice sessions needn't be complicated or time-consuming, especially if pilots and instructors use the Practice Flights associated with this book.

The Practice Flights set initial conditions—aircraft type, location, weather, etc.—for a range of activities, including "non-flying" lessons, such as practicing use of the radios. Loading and starting a Practice Flight is no more difficult than opening a Word document or visiting a website. Restarting a Practice Flight, reviewing a flight track, and other tasks associated with using Flight Simulator as a training aid are equally straightforward.

Starting with only basic controls and a simple PC setup, as suggested in Chapter 1, "About this Book," also can help virtual aviators climb the learning curve at V_Y.

simulators, logging time, and FAA approval

Pilots have accumulated hours at the controls of *Microsoft Flight Simulator* for some 25 years, yet many people in the aviation and virtual aviation communities—including flight instructors—are confused about logging time when using PC-based flight simulations and about which simulators are "FAA approved."

> **note**
> The FAA National Simulator Program (NSP) Web site provides background information, lists of currently approved simulators and training devices, and offers guidance about training devices and simulators for manufacturers, flight schools, and FAA inspectors.

simulators and training devices

To address the second question first, the FAA does not approve simulation software as a stand-alone item. The agency certifies and approves *flight training devices, simulators, Personal Computer Based Aviation Training Devices (PCATDs) and Basic Aviation Training Devices (BATDs)*—combinations of software and hardware (displays, instruments, controls, etc.).

For example, according to FAA Advisory Circular 120-45A, "Airplane Flight Training Device Qualification":

> *An Airplane Simulator is a full size replica of a specific type or make, model, and series airplane cockpit, including the assemblage of equipment and computer software programs necessary to represent the airplane in ground and flight operations, a visual system providing an out-of-the-cockpit view, [and] a force (motion) cueing system which provides cues at least equivalent to that of a three degree of freedom motion system....*

Note that by definition, a "simulator" must emulate a specific aircraft, and it must do so with great fidelity, including a detailed representation of the cockpit.

On the other hand:

> *An Airplane Flight Training Device is a full scale replica of an airplane's instruments, equipment, panels, and controls in an open flight deck area or an enclosed airplane cockpit, including the assemblage of equipment and computer software programs necessary to represent the airplane in ground and flight conditions to the extent of the systems installed in the device; does not require a force (motion) cueing or visual system; is found to meet the criteria outlined in this AC for a specific flight training device level; and in which any flight training event or flight checking event is accomplished.*

Such "flight training devices" range in sophistication from Level 1 (today encompassing mostly obsolete ground training devices) to Level 6 and 7 systems, intended primarily to support training and checks of pilots flying for commercial operators. FTDs may represent specific aircraft or general types of aircraft. The definitions in the advisory circular specify the types of displays, cockpit controls, flight models, and other details required for approval of each level of FTD.

Website blurbs, product reviews, and marketing hype notwithstanding, flight simulation software itself is not "FAA approved" or certified. Simply put, according to the FAA, *Microsoft Flight Simulator* and its software cousins, even when running on a PC equipped with a joystick or yoke and other accessories, are not "simulators," and the software itself cannot be "approved."

A few manufacturers have developed training devices driven by *Microsoft Flight Simulator* (and other products, such as *X-Plane*), some of which have been approved by the FAA and other aviation authorities. Such training devices, which often involve multiple displays and elaborate flight controls and consoles for switches and avionics, cost thousands of dollars (often more than $100,000) and must meet the detailed requirements of AC 120-45A or AC 61-126 "Qualification and Approval of Personal Computer-Based Aviation Training Devices." More on PCATD requirements later.

logging time

The issue of logging time involves two important considerations: How the simulator "flying" is conducted, and how many hours of such time apply to requirements for pilot certificates and ratings.

The FAA regulations answer the logbook question clearly. According to FAR §61.51 "Pilot Logbooks," "A flight simulator or approved flight training device may be used by a person to log instrument time, provided an authorized instructor is present during the simulated flight."

In other words, time spent flying a simulator or FTD counts toward the experience requirements for a certificate or rating, or towards currency requirements, only if the instruction is provided on an approved device and by a rated instructor. Such hours should be logged as "training received in a flight simulator or flight training device from an authorized instructor" (not as "pilot-in-command" or "second in command") and as "instrument time," which does not apply to VFR flying requirements. Simulator time also does not count toward "flight time" (defined as "pilot time that commences when an aircraft moves under its own power for the purpose of flight and ends when the aircraft comes to rest after landing"). Hours accumulated alone at home or while "flying solo" on a flight school's training device (no matter how sophisticated) do not meet the rule's basic requirement.

Contrary to a common misconception, the FAA does not limit the total number of hours a pilot can log on an approved FTD or simulator. However, sections of the FARs that specify the minimum flight experience for pilot certificates and ratings do restrict the simulated hours that a pilot can apply toward those requirements. For example, §61.65 "Instrument rating requirements" notes that a pilot training for an instrument rating may count 20 or 30 hours of simulator time toward the minimums necessary to earn the rating, depending on the type of training program used. A pilot may log as many simulator hours as necessary to learn and practice IFR flying skills, but the FAA recognizes only 20 or 30 of those hours as meeting the regulatory requirements for the rating. This situation isn't unusual. FAA regulations require an applicant for a private pilot certificate to acquire 40 hours of total flight time (under FAR part 61) before taking the practical test for the license. Yet most students today log about 70 hours before they're ready to pass the test.

personal computer based aviation training devices

FAA regulations being regulations, there are, of course, variations on the basic information described above. The only important distinction that relates to the discussion of Flight Simulator in this book concerns a special type of flight training device called a "Personal Computer Based Aviation Training Device" (PCATD) or "Basic Aviation Training Device" (B-ATD).

The need for a special category of flight training devices based on a personal computer developed in the 1990s and was formalized with the publication of AC 61-126 *Qualification and Approval of Personal Computer-Based Aviation Training Devices* in 1997. The advisory circular was the first baby step by the FAA and aviation industry toward formally incorporating PC-based "simulators" into flight training. As such, it sets strict limits on how students and instructors can use PC-based trainers to meet certain training requirements.

Since the publication of the advisory circular that defined the PCATD, the FAA has recognized additional types of aircraft training devices, including the Basic Aircraft Training Device (B-ATD) and Advanced ATD Aircraft Training Device (A-ATD). These changes are described in detail in AC 120-45A and in FAA Order 8700.1 "General Aviation Operations Inspector's Handbook."

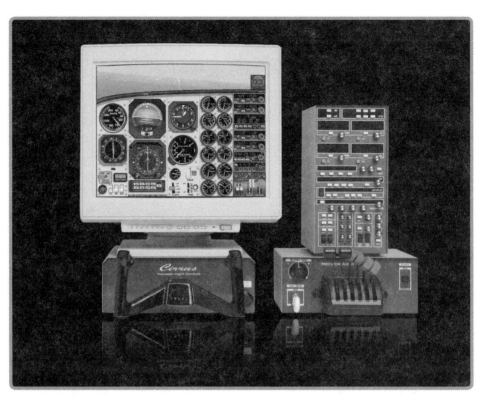

A typical PCATD

Without repeating all of the details in the PCATD advisory circular, here are explanations of key points that seem to cause confusion among virtual aviators, pilots, and instructors:

- The AC defines a PCATD as a device which: "(1) Meets or exceeds the approval criteria [described in the AC]; (2) Functionally provides a training platform for at least the procedural aspects of flight relating to an instrument training curriculum…[and] (3) Has been qualified by the FAA."
- A PCATD may be used to log time toward meeting the requirements for an instrument rating, provided an authorized instructor has presented the instruction and is authorized for use only in an integrated ground and flight instrument training curriculum under [FAR] parts 61 or 141.
- Time logged on a PCATD may be applied in lieu of, and for not more than, 10 hours of the time that ordinarily may be acquired on a flight simulator and credited toward the requirements for an instrument rating. For example, the 20-hour allowance for a flight simulator or an FTD and the 10-hour allowance for a PCATD are not cumulative. If a PCATD is used for the maximum of 10 hours, that 10 hours shall be a part of the 20-hour maximum allowed for a flight simulator or flight training device.
- Time logged on a PCATD applies only toward the requirements to earn an instrument rating. A PCATD cannot be used to conduct a practical test, to meet the currency requirements for IFR, or to meet the experience requirements for any pilot or flight instructor certificate (e.g., sport pilot, recreational pilot, private pilot, commercial pilot, airline transport pilot, or certified flight instructor).
- The FAA allows credit for a little more time if the training is logged on a B-ATD or A-ATD. These devices can be used for instrument proficiency checks and to log 10 hours of the time required for an initial instrument rating. Students can log 2.5 hours of the time required for a private pilot certificate. All such logged simulator time must be supervised by an instructor.

The AC lays out several hardware requirements that a PC-based device must meet to qualify as a PCATD. A few are worth noting here:

- A PCATD must provide some physical controls and may provide some virtual controls.
- Physical controls should be recognizable solely by appearance as to their function and operation. Physical controls eliminate the use of either a keyboard or mouse to control the simulated aircraft.

The basic controls must include:

- A physical, self-centering, displacement yoke or control stick that allows continuous adjustment of pitch and bank.
- Physical, self-centering rudder pedals that allow continuous adjustment of yaw.
- A physical throttle lever or power lever that allows continuous movement from idle to full power settings.
- Physical controls, applicable to the aircraft or family of aircraft replicated, for flaps, propellers, mixtures, pitch trim, communication and navigation radios, clock or timer, landing gear handle, transponder, altimeter, microphone with push to talk switch, carburetor heat, and cowl flaps.

The advisory circular also specifies details of the flight model: "the PCATD must be comparable to the way the training aircraft represented performs and handles…there is no requirement for a PCATD to have control loading to exactly replicate any particular aircraft…a maximum 300-millisecond lag between control inputs and instrument response is required." The AC also specifies details such as the controls provided for an instructor.

A typical PCATD system, the On Top PCATD, currently retails for around $3,000, which includes the On Top simulation software and physical controls and avionics consoles, but not the computer and monitor required to operate the system.

The important point to note about all of the flight training devices discussed above, including PCATDs, is that their sophistication and complexity, while valuable in supporting certain training tasks, also encumber them with disadvantages. They are expensive machines that serve one purpose—simulating an aircraft cockpit—and they must be installed in a fixed location and can support only one or two students simultaneously.

learning, not logging

In the end, parsing regulations about logbooks and training devices isn't productive. The goal of flight training, after all, isn't to log a prescribed number of hours and master a specific set of maneuvers; it is learning to become a competent, safe pilot. Many tools used during flight training (e.g., models, cockpit posters, and PowerPoint presentations) are not "FAA-approved," yet they are still indispensable. Flight Simulator certainly fits into that category.

appropriate level of simulation

The preceding discussion defines the concept of the "appropriate level of simulation," which the FAA and the aviation industry recognize as vital to the development of effective training and overall aviation safety. The certification standards implicitly support the idea that the level of sophistication required to support effective training depends on the purpose of the exercise.

Ironically, the availability of inexpensive, sophisticated simulation software and increasingly powerful PCs blurs the distinction between training devices. What's more, technology often drives the training process. Sometimes such technophilia merely impedes an individual's progress because it overcomplicates a lesson and overburdens a student, who in the pursuit of "realism," is pushed to juggle too many tasks too soon.

More broadly, however, the evolution of PC-based flight simulation can affect the general development of pilot training because pilots and instructors too often focus on what a device *can do* and not on *how creatively it can be used*. And that is the topic for Chapter 3, "Best Practices for Using Flight Simulator."

three

best practices
for using
Flight Simulator

checklist

☐ benefits of using
   Flight Simulator

☐ VFR flying

☐ in the classroom

☐ a graphics tool

☐ using the autopilot

☐ the ATC feature

☐ developing
   "the numbers"

☐ Flight Simulator
   challenges

☐ benefits for
   instructors

☐ uses for practice
   flights

The previous chapter discussed general recommendations for using *Microsoft Flight Simulator* as a training aid and addressed common concerns about including PC-based simulators in flight training programs. Now it's time to focus on specific ideas for using Flight Simulator. Some of the following suggestions are intended primarily for flight instructors, but many ideas should be helpful to anyone who wants to include virtual flying in a training or proficiency program.

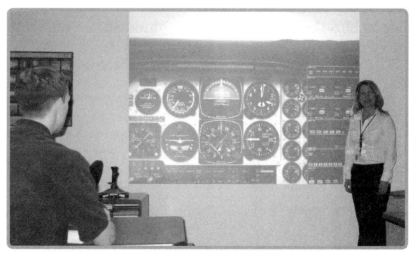

You can use Flight Simulator to make classroom presentations dynamic.

benefits of and advice for using Flight Simulator

All of the suggestions in this chapter focus on the key advantages that Flight Simulator offers over other training devices, especially its:

- Low cost, which encourages unlimited use and regular practice.
- Portability and ease of use, which supports training anywhere a computer is available.
- Flexibility as a training tool, which makes Flight Simulator useful in many settings, including classrooms, briefing cubicles, and the home.
- Role as an aero-experimental laboratory in which to make mistakes so that you don't make them in the air.

In addition, keep the following points in mind when thinking about creative applications for Flight Simulator:

- Divide complex tasks into manageable parts, especially when introducing or learning new procedures. Complete flights from takeoff to touchdown often don't create the best conditions for learning and practice.
- Take advantage of the low-stress environment that Flight Simulator provides and focus on specific tasks to reduce interference and distractions.
- Keep exercises short so that training can occur spontaneously and efficiently almost anywhere and any time.
- Use Flight Simulator as a demonstration tool, not just for "flying."
- Consider how Flight Simulator can help pilots surmount learning plateaus, even if weather, mechanical problems, or other issues cancel flight lessons or "real" simulator sessions.

Flight Simulator and VFR flying skills

Many pilots and instructors think of Flight Simulator primarily as a tool for learning and practicing IFR flying procedures. (Specific applications of Flight Simulator to IFR training are discussed in detail in Chapter 12, "Introduction to the IFR Practice Flights.") However, Flight Simulator can also help VFR pilots master important skills. As noted earlier, the key to the most effective use of Flight Simulator with novice flyers is not to emphasize dexterity with the flight controls required for takeoffs and landing, turns-around-a-point, and other maneuvers, but instead to exploit the visual, interactive potential in Flight Simulator to relate abstract concepts to specific piloting tasks.

For example, a student training for a private pilot certificate must learn about VOR navigation. The background information and diagrams in a text like the *Pilot's Handbook of Aeronautical Knowledge*, however, often prove difficult for a beginner to understand and apply in the cockpit.

Including Flight Simulator in a briefing about VOR navigation can bring the information to life and reduce confusion in the air. Using the autopilot in altitude-hold and heading modes frees the student from the distraction (and frustration) of flying the airplane. It also converts Flight Simulator into a VOR-navigation trainer that supports hands-on practice of all the tasks associated with using VOR navaids, including:

- Finding and identifying appropriate VORs on a chart.
- Obtaining and tuning the appropriate frequencies (Flight Simulator includes information about VOR service volumes; a student who attempts to tune a distant station will learn a valuable lesson).

- Confirming the Morse ID.
- Rotating the OBS to determine an initial line of position.
- Correlating information from the OBS and the heading indicator and then turning the aircraft to track directly to or from a VOR.
- Intercepting and tracking radials, compensating for crosswind components.
- Taking cross-bearings from VORs to fix aircraft position.
- Using DME as an aid to determining aircraft position and for checking groundspeed.

The VFR Practice Flights associated with this book include situations that support such VOR-navigation exercises. These Practice Flights begin in the air, with the aircraft positioned near several navaids. This eliminates the steps normally required to configure Flight Simulator for a VOR lesson.

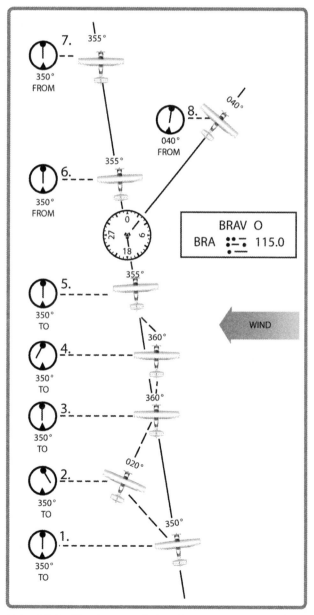

Using Flight Simulator can bring abstract concepts such as tracking a VOR radial to life (from FAA-H-8083-25)

The VFR Practice Flights start you in position to practice specific skills, such as VOR navigation.

Flight Simulator can support many similar VFR lessons and exercises, such as:

- Using sectional charts for pilotage.
- Obtaining and tuning frequencies from the AF/D and charts and then setting up communication and navigation radios appropriately.
- Using VORs for orientation and navigation, including flying along airways and avoiding controlled airspace.
- Visualizing VFR departures and arrivals based on sectional and terminal area charts.
- Operating in and around Class B, C, D, and E airspace.
- Previewing VFR cross-country flights, including diversions caused by weather or mechanical difficulties.
- Visualizing the problems associated with night flight and flight in marginal weather.
- Cockpit orientation.
- Understanding flight and engine instruments.
- Demonstrating the process for setting power in an airplane equipped with a constant-speed propeller.

- Observing the effects of flying at best-rate and best-angle of climb speeds; how weight and density altitude affect performance, and so on.
- Operating cockpit controls and avionics.
- Practicing use of checklists and cockpit flows.
- Learning about and practicing cockpit resource management.
- Rehearsing ground operations.

None of these exercises requires that the student actually fly Flight Simulator, and all of them can be introduced to a group of students in a classroom or to an individual pilot during a one-on-one briefing. Students who have Flight Simulator on a computer at home can repeat the exercises whenever they have a few minutes available. Focusing on specific tasks and concepts rather than emphasizing use of the flight controls also helps reduce the potential for negative transfer from desk-flying to the real cockpit, especially with inexperienced pilots.

climbing over learning plateaus

Applying Flight Simulator to the types of exercises outlined in the previous sections becomes especially helpful when students encounter learning plateaus and when weather or other issues interrupt flight lessons.

Suppose a student is primed for a solo cross-country trip but must postpone it several times due to weather. Instead of leaving skills like VOR navigation to atrophy, a student can use Flight Simulator to rehearse key parts of the flight. Again, use of the flight controls to hand-fly the entire trip is not required for training benefits to accrue. Just being in "pilot mode" and practicing specific skills, such as a maintaining a flight log, can be a big help and a confidence-builder.

Flight Simulator can help VFR pilots understand the challenges associated with flying in marginal weather.

An instructor can also use the real-world weather feature in Flight Simulator to show a student why launching on a cross-country trip in marginal conditions isn't a good idea. On an MVFR day, start Flight Simulator along a planned route. The student can see how difficult it becomes to track position and identify checkpoints at low altitude in limited visibility. Flight Simulator can also help an instructor vividly demonstrate being squeezed between weather, terrain, and controlled airspace; how strong crosswind components complicate navigation; the sudden appearance of airports in low visibility; and the limitations on VOR reception at low altitude. All of these exercises can be conducted as demonstrations, with the autopilot handling basic flying chores.

more information about using Flight Simulator for VFR lessons

Chapter 10, "Introduction to VFR Practice Flights," provides detailed information about the Practice Flights created specifically for students and VFR pilots. Many of the VFR Practice Flights can be used to jump-start demonstrations such as those suggested above.

using Flight Simulator in the classroom

The preceding discussion advocates using Flight Simulator as a demonstration tool, not just as a virtual cockpit. In fact, Flight Simulator can easily become an interactive, dynamic substitute for or complement to a whiteboard, PowerPoint presentation, poster, aircraft component, or model. Instead of turning by habit to those teaching aids, instructors should consider how they might put Flight Simulator to work in situations where they need to illustrate concepts and procedures. Using Flight Simulator as a demonstration tool allows any flight school or independent instructor with access to a PC to include vivid, interactive elements in presentations and briefings. I also frequently include Flight Simulator in presentations to pilots, schools, and community groups.

Projecting Flight Simulator cockpits makes classroom presentations dynamic.

Projectors and large monitors make it easy include Flight Simulator in lessons for groups of students. For example, an instructor working with one or two students can gather everyone around a standard PC monitor. When teaching a larger group, using Flight Simulator with a computer projector or displaying it on large monitor lets everyone see what's going on.

An instructor can use Flight Simulator in this way to:

- Show flight and engine instruments in action and in context. For example, the instruments in a Flight Simulator cockpit can substitute for a cumbersome display model of a VOR indicator, HSI, or other gauge.
- Demonstrate the process of tuning, identifying, and determining position with VOR and ADF systems.
- Jump outside the airplane in Spot Plane view to describe the important features of different aircraft and to observe the control surfaces in action.
- Taxi around an airport to learn about signs and markings. Use wide view to "see over the panel" and top-down view to see the entire airport layout.
- Provide students with an appropriate VFR or IFR chart or approach plate, and have them observe and describe the indications on the navigation instruments as the aircraft "flies" along a route or segments of an approach.
- Demonstrate specific procedures, such as holding pattern entries.
- Change weather settings and demonstrate what the elements described in FAR §91.175(c) look like as an aircraft arrives at the minimum altitude or MAP specified for an instrument approach.
- Set the time of day to just before dawn or just after sunset, and as the aircraft flies toward a runway, observe the functions and effects of lighting systems, including VASI and PAPI approach aids.

Throughout such demonstrations, instructors should take advantage of the pause feature in Flight Simulator (press the P key to freeze the simulation; press it again to resume flying) at key points, as if lingering over a PowerPoint slide or whiteboard diagram to discuss important details.

When using Flight Simulator as demonstration tool, especially for a group of students, the autopilot can handle basic flying chores and help the instructor concentrate on teaching. If the class includes students who are adept at flying Flight Simulator, they can take the controls occasionally. Using such copilots helps keep them engaged in the discussion.

Flight Simulator as a tool for the graphically challenged

Pictures enhance briefings and classroom sessions, but many instructors (including myself) are, to put it charitably, best appreciated as abstract artists. Flight Simulator can come to the rescue as a source of detailed depictions of instrument panels, airport environments, and other elements that enhance PowerPoint slides and other teaching materials. (Microsoft has published guidelines for the use of screen shots of its products, including *Microsoft Flight Simulator*, on the "Use of Microsoft Copyrighted Content" page of the corporate website.)

capturing a screen

To take a snapshot of any view at any time when Flight Simulator is running:

1. Set up Flight Simulator to show the instrument panel, outside view, map view, or other scene that you want to capture, and, if necessary, pause the simulation.
2. Press the PRINT SCREEN key. Whatever is on the screen is copied as a bitmap image (.bmp) to the Clipboard (a temporary holding area for text and graphics in Windows).
3. Next, switch to Paintbrush (the built-in graphics editing tool in Windows) or another picture-editing tool, and then copy the captured image into that application (usually by clicking the **Paste** command on the **Edit** menu).

4. Save the image, preferably as a JPEG (.jpg) file.
5. Import that JPEG image to a PowerPoint slide or graphics editing program and add arrows, explanatory text, and other embellishments.

> **note** If you use Flight Simulator X, you can "take a photo" at any time by pressing the **V** key. For more information, see the topic "Photos" in the Learning Center.

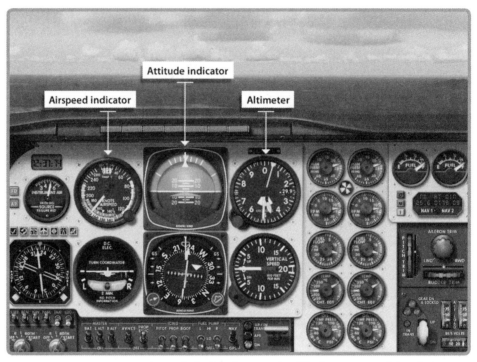

You can use screen captures from Flight Simulator to simplify the creation of detailed illustrations.

You can capture images of the instrument panel at key points in a flight, take pictures from Spot Plane view to point out features of an airplane, and even snap full-screen, in-flight views of landmarks and airports to compare with portions of a sectional or other chart.

You can also use a screen-capture utility such as SnagIt to take screen snapshots in Flight Simulator. Programs like SnagIt (published by TechSmith Corporation) offer more flexibility in capturing, editing, and saving images than the basic utility of the PRINT SCREEN key in Windows.

using the autopilot in Flight Simulator

Yogi Berra, (quoted earlier in Chapter 2) also said, "You can observe a lot just by watchin'." Many flight schools incorporate that wisdom into their training programs by encouraging students to fly in pairs. One trainee sits in the left seat next to the instructor and gets hands-on practice at the controls. The other student rides in back to watch, listen, and learn. These back-seat flyers don't have to divide their attention between flying and learning, so they can concentrate on following procedures, using charts, interpreting the instruments, listening to ATC instructions, and learning from the other student's mistakes.

Autopilot in the Cessna 172

Virtual aviators can employ the autopilot in Flight Simulator to accomplish the same goal. Debates about the appropriate use of an autopilot during flight training notwithstanding, using the virtual autopilot is key to getting the maximum benefits from using Flight Simulator effectively as a teaching aid. Instructors should encourage students to use the heading bug and autopilot altitude controls to "fly" the airplane until they thoroughly understand the concepts underlying a particular scenario. As students gain proficiency with Flight Simulator and experience flying real aircraft, instructors can turn off the autopilot and require students to juggle more tasks simultaneously. The same advice applies to pilots who use Flight Simulator to practice procedures at home.

the ATC feature in Flight Simulator

The ATC feature in Flight Simulator is not as comprehensive as the real air traffic control system, and it has several limitations that affect its utility as a training tool. For example, virtual aviators interact with controllers through pop-up windows that display context-sensitive menus. ATC transmissions and pilot responses play through the computer's speakers. Unfortunately, the virtual ATC doesn't respond to voice commands. (We're still a few years away from adding speech recognition to all of the other burdens—calculating flight models, animating instruments, generating weather, displaying scenery, etc.—of simulating flight on a single PC.)

The phraseology employed by the Flight Simulator ATC system doesn't always precisely mimic the recommendations in the AIM and Pilot/Controller Glossary. For example, the voices for ATIS and ASOS broadcasts spell out identifiers instead of pronouncing facility names (doing otherwise would have required recording tens of thousands of place names).

Nevertheless, an instructor can still use the ATC feature to help students develop a sense of communication flow during a typical flight and to add realism to practice sessions. Even if a virtual training flight doesn't include interacting with ATC, conversations between the computer-generated traffic and controllers add the equivalent of a movie score to the experience.

Interactions with ATC in Flight Simulator rely on pop-up menus.

That said, ATC can be a distraction when focusing on specific tasks. I usually turn it off when teaching with Flight Simulator or flying the Practice Flights associated with this book.

For more information about the ATC feature, see the "Air Traffic Control" section in the Flight Simulator Learning Center.

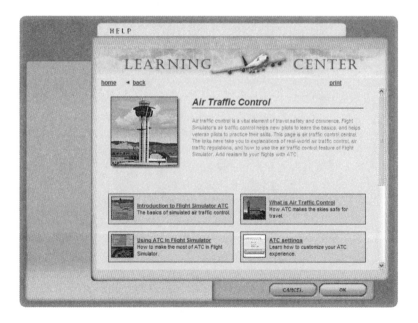

using Flight Simulator to develop "the numbers" for aircraft

Experienced pilots stay ahead of the airplane and reduce their overall workload by learning "the numbers" for each aircraft they fly. Applying the principle that "Pitch + Configuration + Power = Performance," they experiment with a range of pitch attitudes, flap and landing gear configurations (on aircraft with retractable landing gear), and power settings and then create reference tables that help them quickly establish stable flight for common situations.

As an example, consider this configuration in a Cessna 172: Pitch one bar width below the horizon (wings of the miniature airplane on the attitude indicator rest just below the horizon line), flaps set to 10 degrees and power set to 2000 rpm. This will generally put the airplane on a smooth, stable descent down a typical ILS glideslope—descending at roughly 500 fpm and 90 knots.

Configuration tables for the aircraft used in the Practice Flights are included in Chapter 8, "Flying the Aircraft Used in the Practice Flights."

A configuration for a smooth descent along a glideslope in a Cessna 172.

Flight Simulator is an excellent tool for teaching and practicing this important concept. A student and instructor can use the autopilot to establish stable configurations for an airplane and then experiment by changing power, flap, and gear settings to observe how the autopilot adjusts other variables. (By the way, this technique works in real airplanes, too.)

For example, while using the autopilot in ALT and HDG mode:

1. Set power at a normal cruise setting. Reduce power in 100 rpm (or 1" of MP) increments. Allow time for the airplane to stabilize after each change in power, and then observe the effects on pitch attitude and indicated airspeed. Note the pitch attitude, power, and airspeed after each change.
2. Starting at an appropriate airspeed, add flaps in stages. Observe and note the same effects.
3. Establish a descent and then level off. Note the power required to maintain airspeed in different configurations (useful when leveling off before joining the traffic pattern or when leveling at the MDA on an instrument approach).

best
practices
for using
flight
simulator

Using the autopilot initially during such drills eliminates pilot-induced oscillations and helps students observe how changing one variable affects the others. After a student has learned "the numbers" that will stabilize the aircraft for certain flight conditions, the instructor can turn off the autopilot. The student can then apply that knowledge in practice exercises such as entering and flying a traffic pattern, setting up slow flight, and establishing constant-airspeed and constant-rate descents and climbs.

Although the numbers developed with Flight Simulator probably won't exactly match those in a specific airplane under all combinations of gross weight and ambient conditions (they will be close in similar classes of aircraft), such exercises demonstrate important principles and allow a student to practice setting up the airplane in many configurations.

Flight Simulator challenges

Because pilots are a competitive bunch, most aviation magazines publish quizzes to test aviators' knowledge. Instructors constantly quiz students during briefings and in flight. Flight Simulator can add interactive elements to such exercises in several ways.

For instance, instead of poring over and "chair-flying" an instrument approach plate, an instructor and student can fly the procedure virtually in Flight Simulator and discuss how to set up the avionics, configure the airplane, navigate between key points, and track position throughout the scenario.

VFR students and instructors can also use Flight Simulator to rehearse many situations, such as arriving at an unfamiliar airport, diverting from the original destination on a cross-country flight, and maneuvering through or around controlled airspace.

It takes only a few moments to set up Flight Simulator for such challenges. For more information about creating, saving, and using custom Practice Flights, see Chapter 14, "Creating Your Own Practice Flights."

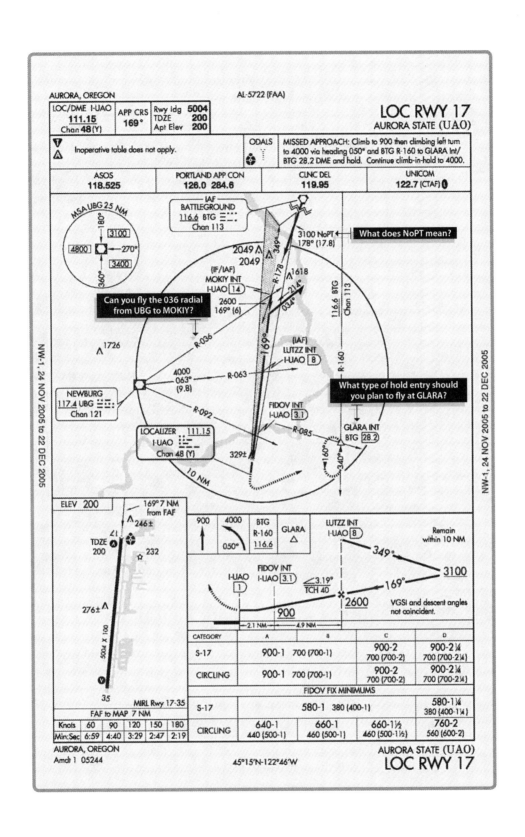

benefits for instructors who use Flight Simulator

Using Flight Simulator as a complement to ground instruction has an additional benefit beyond the advantages discussed earlier in this chapter. Flight Simulator is a great tool for testing students' working knowledge. With all the study tools available today, it has become easy for students to parrot correct answers to favorite tricky questions.

But putting students in front of Flight Simulator and giving them an exercise involving a complex task, such as VOR navigation or diverting during a cross-country flight, soon shows if they can apply theory to solve practical problems in real time. Most importantly, instructors and students can find out if additional study is required before they get into the airplane. Again, such exercises need not focus on hand-flying. The goal is to see if a student can follow the correct procedures and use all available resources to solve the problem.

other uses for the practice flights

The Practice Flights associated with this book aren't just for flying. They set initial conditions for a variety of situations and lessons that give instructors and students a head start if they want to use Flight Simulator for demonstrations and other activities described throughout the book.

For example, the Basic Attitude Instrument Flying IFR Practice Flights begin in the air with the autopilot on. Although intended primarily to support lessons related to developing an instrument scan and for flying basic IFR patterns, these Practice Flights can also support other activities, such as showing the cockpit instruments to beginning VFR students and giving classroom presentations on the functionality of the flight controls.

four

Flight Simulator essentials

checklist

☐ learning resources

☐ kneeboard

☐ joystick

☐ using the mouse

☐ looking around

☐ view modes

☐ autopilot

☐ using "flights"

☐ the map

☐ instant replay and videos

☐ GPS

☐ weather

☐ systems failures

☐ customizing Flight Simulator

Much of the reluctance—sometimes resistance—to using *Microsoft Flight Simulator* in flight training programs stems from frustration with or fear of computers. For many pilots and flight instructors, a PC is far more intimidating than an airplane. But anyone with enough computer skills to check email, open a Word document, or surf the Web knows enough about PCs to put Flight Simulator to effective use as a training aid.

To repeat a key point from Chapter 1, "About this Book," Flight Simulator is a *simulation* that accurately reproduces the experience of flying an aircraft. This means that the aviation knowledge and skill pilots and instructors already have will transfer almost directly to virtual flying. More importantly, aviators do not need to memorize lists of arcane keyboard commands or know anything about programming. This applies as much to the Practice Flights in this book as it does to creating individualized custom situations.

Differences between Flight Simulator 2004 and Flight Simulator X mean that some menus and other details such as launching Flights aren't the same (for more information about important disparities, see Chapter 6, "Differences Between Flight Simulator 2004 and Flight Simulator X"). Nevertheless, the fundamentals of using Flight Simulator, just like basic flying skills, apply to both versions. Switching between versions is akin to going from an airplane with a carburetor to one with a fuel-injected engine.

*You can always get help—press the **ALT** key to display the menu bar.*

learning resources and help

As you become acquainted with Flight Simulator or want to use specific features, such as Flight Analysis, that are important components of the Practice Flights, remember that the Flight Simulator Learning Center contains hundreds of pages of text, videos, pictures, illustrations, and links to additional information about the simulation. The Learning Center is always available because it is an integral part of Flight Simulator. Think of it as a combination Pilot's Operating Handbook and *Pilot's Handbook of Aeronautical Knowledge* for Flight Simulator.

You can browse the Learning Center like a website—it is composed in HTML, the language of the web—and use familiar skills, such as clicking links, to find more information.

To find specific information in the Learning Center, you can use the illustrated Key Topics screen, which takes you directly to overviews of essential information about the most important features in Flight Simulator.

You can open the Learning Center from the Help menu.

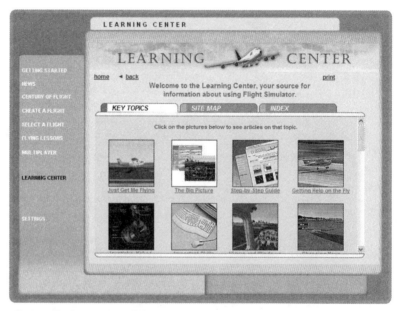

The Key Topics section of the Learning Center is the gateway to information about important features.

You can also find specific topics in the Index and definitions of terms in the Glossary.

The Site Map for the Learning Center provides another way to find the information you need.

I have also created a Microsoft PowerPoint presentation, *Flight Simulator Fundamentals*, included on the CD that accompanies this book. If you're new to Flight Simulator or would like to pick up some tips about using it more effectively, take a few minutes to review that presentation. You don't need PowerPoint to view the presentation.

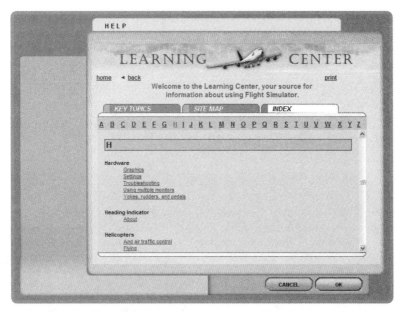

Use the Index tab in the Learning Center to jump directly to specific information.

the kneeboard

Flight Simulator includes a virtual kneeboard modeled after the handy organizer and notepad that many pilots use in the cockpit.

The Flight Simulator Kneeboard includes six pages:

- **Briefing:** Displays the briefing for the Flight, if there is one.
- **Radio:** Logs the last 10 Air Traffic Control radio transmissions to your aircraft.
- **Navigation Log:** Provides a list of waypoints, headings, and other information for a flight plan created using the Flight Planner.
- **Key Commands:** Provides a list of keyboard commands.
- **Checklists:** Lists step-by-step procedures for the aircraft you're flying.
- **Reference:** Lists recommended speeds for the aircraft you are flying: how fast to fly during each phase of flight, and what the limits are.

You can always display or hide the Flight Simulator Kneeboard by pressing a button on a joystick or yoke, clicking the Kneeboard icon on any instrument panel, or by pressing the F10 key.

default joystick/yoke/throttle controls

Most joysticks available today include all of the basic controls you need to fly Flight Simulator, regardless of the type of aircraft you prefer.

For example, a typical Logitech joystick has a built-in throttle, a "hat" switch for looking around, buttons to extend and retract the landing gear and flaps, and other knobs, buttons, and levers for functions such as elevator trim and brakes. The joystick also twists to provide rudder control .

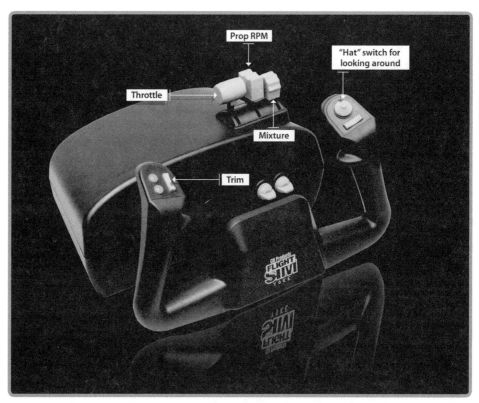

CH products Flight Sim Yoke

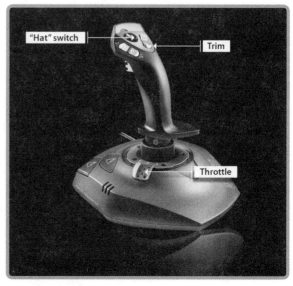

Logitech joystick

The CH Products Flight Sim Yoke also includes all these features (except for rudder control).

Each manufacturer numbers the buttons and other controls on its devices differently, but by default, Flight Simulator assigns the following functions to these joystick/yoke controls:

action	command
apply/release brakes	button 1 (trigger)
cycle view categories (cockpit, tower, track, spot)	button 2
elevator trim down	button 3
elevator trim up	button 4
extend flaps in increments	button 5
retract flaps in increments	button 6
display/hide kneeboard	button 7
landing gear up/down	button 8
bank left (ailerons)	move stick left
bank right (ailerons)	move stick right
pitch down (elevator)	move stick forward
pitch up (elevator)	move stick backward
yaw left (rudder)	twist stick left
yaw right (rudder)	twist stick right
look straight ahead (12 o'clock)	move hat switch up
look left (9 o'clock)	move hat switch left
look right (3 o'clock)	move hat switch right
look back (6 o'clock)	move hat switch down
look ahead and to the left (10 o'clock)	move hat switch up/left
look ahead and to the right (2 o'clock)	move hat switch up/right
look back and to left (8 o'clock)	move hat switch down/left
look back and to the right (4 o'clock)	move hat switch down/right

To sort out which button or control on your joystick or yoke does what, experiment and consult the manufacturer's guide or website. Most manufacturers have created "profiles," or customized control configurations, for *Microsoft Flight Simulator*. You can download these profiles from the manufacturers' websites.

Remember, you can reassign actions to any joystick or yoke button or control. For more information, see the topics "Using a Joystick" and "Customizing Joystick Assignments" in the Learning Center. You can also calibrate and adjust the sensitivity of the main joystick or yoke controls. For more information, see the topic "Adjusting Joystick Sensitivity" in the Learning Center.

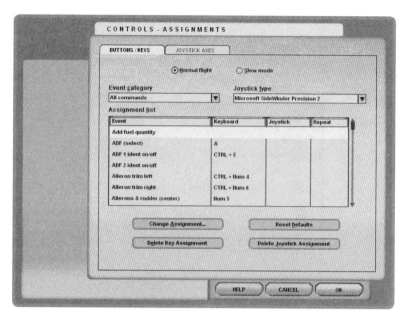

Controls-Assignments dialog box

using the mouse in the cockpit

If your computer is equipped with just a basic joystick and a mouse, you don't need a keyboard during a typical flight in Flight Simulator. In fact, a keyboard is not required even to start Flight Simulator and load a Flight. As for what to do after you climb into the virtual cockpit, just remember the key concept already mentioned in Chapter 2:

"While using Flight Simulator, think of the mouse as an extension of your hand…"

For example, to tune a radio, point to the part of the frequency you want to change, and then slowly roll the mouse wheel forward to increase the digits or back to decrease the digits.

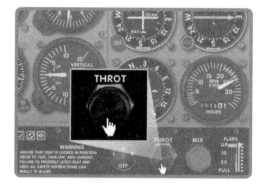

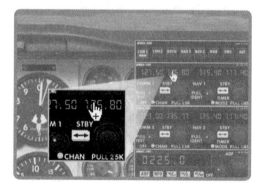

Almost any real cockpit control that you would operate with your hand can also be changed or adjusted in Flight Simulator with the mouse. For example, you can drag the mouse, roll the mouse wheel or click buttons to:

- Adjust the throttle, propeller, and mixture controls.
- Change heading with the heading bug (when the autopilot is ON and in HDG mode).
- Extend and retract the flaps and landing gear.
- Adjust elevator, rudder, and aileron trim.
- Change altitude and rate of climb or descent with the autopilot controls (when the autopilot is ON and ALT mode is selected).
- Tune communication and navigation radios, swap standby and active frequencies, and select radios to monitor on the audio panel.
- Select courses on the VOR/ILS indicators and HSI.
- Adjust the altimeter setting.
- Turn lights on and off.
- Display and hide the popup windows for the avionics, Kneeboard, map, GPS, and additional aircraft controls.

> **note** Flight Simulator X re-introduces the "mouse as yoke" feature from early version of *Microsoft Flight Simulator*. When you select this option, you can use the mouse as if it were a joystick or yoke.

looking around and changing views

Often the most frustrating limitation associated with using a PC-based simulation is the restricted view offered through the "window" of a typical computer monitor. You can't eliminate this constraint without setting up a wrap-around view, but you can make it less annoying by learning to use the basic controls for "looking around" and changing views in Flight Simulator.

normal cockpit view

By default, the Practice Flights associated with this book start in normal cockpit view, which features a straight-ahead orientation out the window with an instrument panel displayed in the lower part of the screen.

Normal cockpit view

While in cockpit view, you can "turn your head" by moving the "hat switch" on the joystick or yoke. Press and hold the switch in the direction you want to look. When you're ready to look straight ahead again, release the switch.

When you are in virtual cockpit or "3D" view (described below), you can also press the "hat switch" to pan—slowly "rotate your head."

To "raise your seat" to get a better view over the instrument panel, press SHIFT+ENTER repeatedly until you're sitting as "high" as you like. To "lower your seat," press SHIFT+BACKSPACE.

To zoom in and out in cockpit view (or any view mode described below), press the PLUS SIGN and MINUS SIGN keys. As a general rule, it is best to avoid zooming in and out.

You can make other adjustments while in different view modes. For details, see the topic "Using Views and Windows" in the Learning Center.

w: hide panel view

In my opinion, the best option for "taking a peek" over the instrument panel while practicing visual maneuvers, during the initial stages of takeoff, and in the final seconds of an approach and landing is "wide-screen" view or "hide panel" view.

To remove all but the essential cockpit instruments, press the W key once. Press W again to clear all the instruments. To restore the instrument panel, continue the cycle by pressing W once more.

This view is so handy that I often assign it to button 2 on a joystick or yoke to replace the default function of that button (cycling through the main view modes in Flight Simulator). Assigning this view to the joystick button means I don't need access to the keyboard when I'm flying or demonstrating to students.

view modes in Flight Simulator

The default view from the pilot's seat works well for almost all applications of Flight Simulator as a training aid. But sometimes you may want to take advantage of other view modes in Flight Simulator, which include:

- Cockpit view (described above)
- Virtual Cockpit view
- Tower view
- Spot plane view
- Top-down view

By default, you can cycle through these view modes by pressing button 2 on your joystick or yoke or by pressing the S key or SHIFT+S on the keyboard.

When you switch views, a text message identifying the current view mode briefly appears in the upper-right corner of the screen.

Spot plane view

All of these view modes are described in detail in the topic "Using Views and Windows" in the Learning Center. A few key points, however, are worth reviewing here.

virtual cockpit

The virtual cockpit or 3D view became practical for flying in Flight Simulator 2004 because the instruments react as the airplane flies. You can also operate many cockpit controls with the mouse, including the radios and autopilot.

Virtual cockpit view

This view provides a wider, deeper perspective of the world, useful when demonstrating and practicing visual maneuvers and general concepts. The gauges can be difficult to read, however, so I don't recommend switching to virtual cockpit view to polish IFR skills.

tower view

The so-called tower view in Flight Simulator is fixed at a spot on your departure airport, regardless of whether that airport actually has a control tower. Because the airplane quickly flies out of range (unless you stay in the traffic pattern), I rarely switch to tower view, especially when using Flight Simulator as a training aid.

Tower view

spot plane (outside view)

Spot plane or outside view

Spot plane view shows your aircraft from outside, as if from the cockpit of a wingman flying in close formation. It's fun to see your airplane skimming over the clouds or cruising over interesting scenery, but this view has limited value as a training aid. You can switch to spot plane view to show how control surfaces move as you wiggle the flight controls. Or use it to show the outside view of a landing or other maneuver when when reviewing a flight with the Instant Replay feature.

top-down view

The top-down view provides a bird's-eye perspective on your aircraft. This view is most useful when you're on the ground at an airport and want to plan a route to the runway.

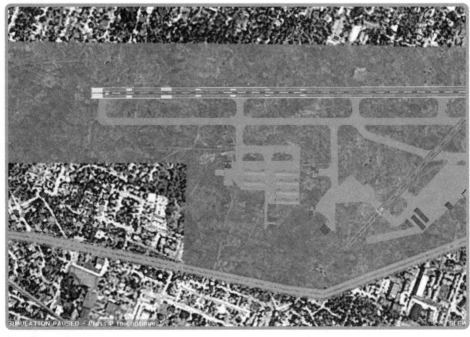

Top down view

the Flight Simulator autopilot

The autopilot installed in the general aviation aircraft in Flight Simulator emulates a typical light-aircraft autopilot. In the Baron, Mooney, and other high-performance aircraft, it includes a flight director. You can learn all about the autopilot in the topic "Using an Autopilot" in the Learning Center.

The Cessna 172 and the Beechcraft Baron 58 used for the Practice Flights associated with this book are equipped with autopilots that can:

- Maintain a selected heading.
- Maintain a selected altitude.
- Maintain a selected rate of climb or descent (assuming sufficient available engine power).
- Track a VOR radial.
- Track a localizer or localizer back course.
- Track the localizer and glideslope of an Instrument Landing System (ILS).
- Track a GPS course.
- Hold the wings level so that the airplane does not turn.
- Maintain the aircraft's current pitch attitude.

using the autopilot with the practice flights

Most of the Practice Flights start with the autopilot ON with HDG and ALT modes active. This configuration helps the situations begin in a stable configuration and simplifies control of the aircraft for normal maneuvering.

To turn the aircraft with the autopilot on, point to the heading knob for the heading indicator or HSI and roll the mouse wheel forward or back until the bug is set to the heading you want to fly.

To climb or descend with the autopilot in ALT mode, point to the altitude display on the autopilot and roll the mouse wheel forward or back to change the digits. You can also use the mouse to adjust the rate of climb or descent in the vertical speed display on the autopilot.

To turn the autopilot on or off:

- Click the AP button on the autopilot.

-or-

- Press the Z key

-or-

- Click the AP annunciator on the instrument panel.

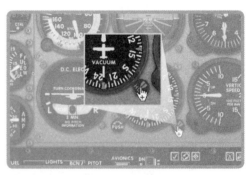

Use the mouse to move the heading bug on the heading indicator or HSI.

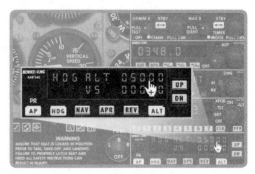

Use the mouse to select an altitude or vertical speed on the autopilot.

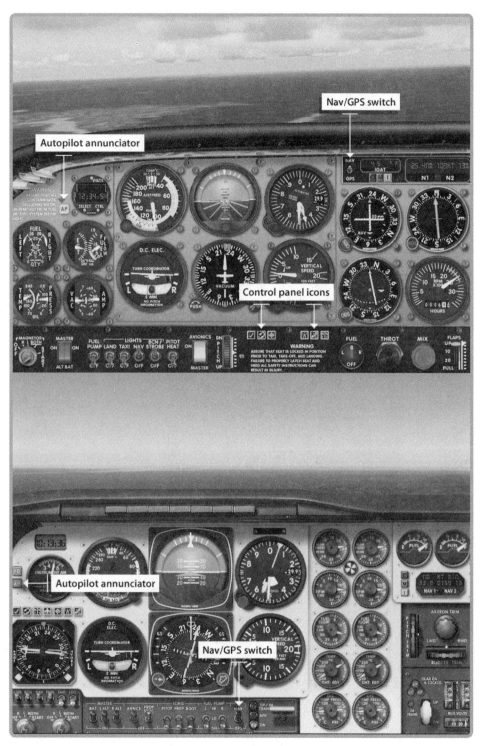

You can use the mouse to operate autopilot controls on the main instrument panel.

Two other autopilot modes are handy when using Flight Simulator as a training aid. In NAV mode the autopilot tracks the VOR or localizer signal from the number one navigation receiver. If the GPS is driving the number one VOR indicator or HSI, NAV mode tracks the GPS course.

To control which signal drives the NAV function of the autopilot, use the mouse to flip NAV/GPS switch on the instrument panel.

using "flights"

Flights get you started quickly in a specific aircraft at a particular location, with weather, views, and other conditions already set up. Chapter 7, "About the Practice Flights," describes the Flights feature and the Practice Flights associated with this book in more detail. For now, just note that using the Flights feature in Flight Simulator is no more complicated than opening and saving a document.

You can also learn more about the Flights feature by reviewing "All About Flights" in the Learning Center. For quick reference, however, here is the essential information you need to start, reset, and end a Flight.

starting a flight

You can start a Flight from the main Flight Simulator screen or by clicking the **Select Flight** command on the **Flights** menu.

Either action opens the **Select a Flight** dialog box from which you can choose a specific Flight.

Use the commands on the Flights menu to select (start), reset, and save Flights.

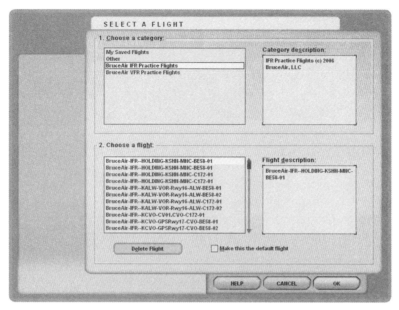

Use the options in the Select A Flight dialog box to start a Flight.

resetting (restarting) a flight

At any point during a Flight, you can reset the aircraft's initial position and restore the original configuration of the weather, avionics, and other starting conditions defined in the Flight.

If you have a keyboard handy, the quickest way to restart a Flight is to press the CTRL+; (semicolon) keys.

You can also restart a Flight by choosing the **Reset Flight** command on the **Flights** menu.

ending a flight

- On the **Flights** menu, choose the **End Flight** command

-or-

- Press ESC

-or-

- Load a new Flight

the Flight Simulator map

The Flight Simulator Map resembles the moving maps now common in aircraft cockpits—with one important exception. It doesn't "move." That is, you can't use the map to watch the aircraft you're flying move in real time. When you display the map, Flight Simulator pauses. (To track your progress on a moving map as you fly, display the GPS.)

The map is very handy when using Flight Simulator as a training aid. It provides a quick way to reposition and re-orient the aircraft, to confirm the aircraft's position before starting a Flight, and to get information about airports and navaids.

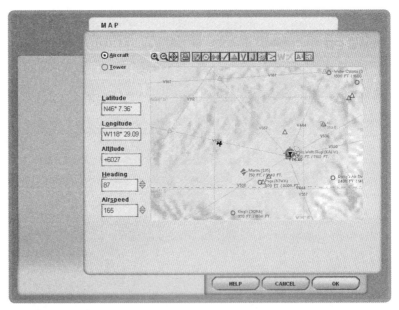

The Flight Simulator map

For example, in map view, you can quickly drag the airplane to a new location, and then set a specific heading, altitude, and airspeed that will apply when you return to the cockpit.

To display the map, click the map icon on the instrument panel, or on the **World** menu, click **Map**.

You can use controls on the map screen to adjust the zoom level of the map and display or hide such items as airports, and VORs, airways.

For more information, see the topic "Using the Map" in the Learning Center.

To display the map, click the map icon on the instrument panel.

instant replay and flight videos

Flight Simulator includes an instant reply function that works much like the reviews of the action featured during sporting events. Instant replay automatically records what happens in Flight Simulator as soon as you start a Flight, but the record is erased when you end a flight, load another flight, or exit Flight Simulator. If you want to admire that squeaker landing, do so right away.

The Flight Video feature in Flight Simulator records and saves the same information as the instant replay feature so that you can review it at any time. Note, however, that this function does not create true videos (such as AVI or MPEG recordings). Instead, the flight video feature works more like a flight data recorder installed in airliners. It continuously stores the aircraft's position, attitude, speed, and other information in a file. When you "play back" a flight video, Flight Simulator runs that information through the simulation engine to re-create the action on screen. In other words, you can play back flight videos only through Flight Simulator. (You can send the .vid file that contains the information saved in a Flight Simulator "video" to anyone who also has Flight Simulator for playback, but .vid files for long flights can grow large, and the other users must know where to save the .vid file on their computers and how to use the flight video feature in Flight Simulator.)

For more information about these features, see the "Instant Replay" and "Flight Videos" topics in the Learning Center.

making real videos

It is possible to capture video from a Flight Simulator if your computer has a video card that supports two outputs, one for the monitor, and another for a second device. You can connect a video recorder, such as a camcorder, to the video card and capture the action on the screen. The procedure to follow and the quality of the resulting recording depend on the capabilities of the video card in your computer and the type of recorder you use.

You can also buy software, such as Camtasia Studio from TechSmith, to capture video as you use your computer. To learn more about video-capture utilities, search for reviews and product information on websites such as PC Magazine and CNET.com, and on websites for Flight Simulator enthusiasts. Note, however, that such utilities may not work well with graphic-intensive products such as Flight Simulator and games, especially on less powerful computer systems.

flight analysis

Flight Simulator includes a Flight Analysis feature that automatically records your altitude, heading, speed, and other information as you fly. It displays the aircraft's track over the ground and other information. The controls on the Flight Analysis screen work like buttons on a television remote so that you can review the flight and accelerate, slow down, or stop the playback. You can adjust the zoom level of the map and display or hide such items as airports, and VORs, airways.

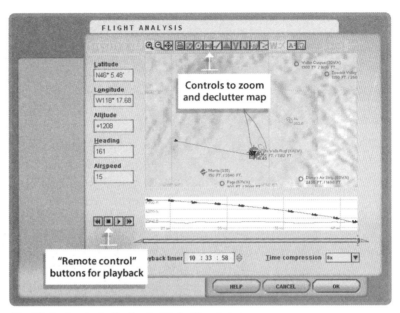

The Flight Analysis display in Flight Simulator.

To display the Flight Analysis screen at any time during or after a flight, on the **Options** menu, choose **Flight Analysis**.

Like the Instant Replay feature, Flight analysis is available until you end flight, load another flight, or exit Flight Simulator. For more information, see the topic "Flight Analysis" in the Learning Center.

the Flight Simulator GPS

Several aircraft in Flight Simulator are equipped with a Garmin GPS 500-series panel-mount GPS receiver, which you can use for IFR operations, including GPS approaches. Other aircraft include a Garmin 295 portable receiver, an excellent aid to VFR navigation.

GPS 500 window, C172

GPS 295 window, Piper Cub

The virtual GPS receivers in Flight Simulator do not replicate all the features of the real units, but most basic navigation functions and displays are similar to those on the actual devices. To learn more about the Garmin GPS units in Flight Simulator:

- Review the topic "Using the GPS" in the Flight Simulator Learning Center.
- Download the complete Garmin 500 series or Garmin 295 Pilot's Guide from the Garmin website. The Quick Reference Guides for those units are also available on the Garmin website.

To learn more about how GPS works and how to use it during VFR and IFR operations, see:

- The articles "About GPS" and "GPS Guide for Beginners," available on the Garmin website.
- AOPA ASF Safety Advisor "GPS Technology."
- AIM, Section 1-1-19.

weather

Flight Simulator includes a weather system that can emulate current conditions around the world or generate custom weather that you define. Flight Simulator also includes "weather themes" that make it easy to specify general conditions such as clear skies and light winds or dark and stormy nights.

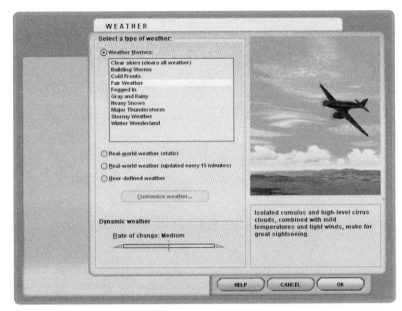

Use the options in the Weather dialog box to control the weather in Flight Simulator.

The ability to create low clouds and reduced visibility comes in handy when you are trying to learn how weather affects flying. For example, you can set up VFR or IFR weather minimums and practice transitioning from instrument to visual references as you break out on an approach.

Many of the Practice Flights associated with this book are available in VMC and IMC versions (visual and instrument meteorological conditions). This way, you can fly first with visual cues, and then without them.

For more information about the weather system in Flight Simulator, see the topic "Introduction to Flight Simulator Weather" in the Learning Center.

systems failures

The greatest benefit simulation provides is the ability to create abnormal conditions and emergencies in safe environment. Flight Simulator can reproduce many types of failures that affect individual instruments, aircraft systems, avionics, and engines.

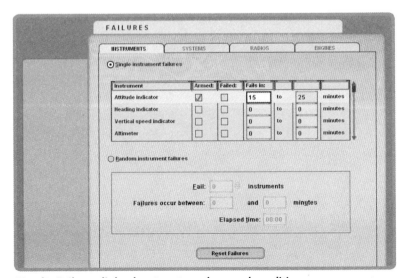

Use the Failures dialog box to set up abnormal conditions.

The system failures in Flight Simulator fill an important gap in the training available at a typical flight school. For example, an instructor can "fail" a gyro instrument in a light aircraft only by reaching over and covering it. The student immediately knows which instrument has failed and isn't distracted by the erroneous indications that would accompany an actual failure.

The gyro and vacuum system failures in Flight Simulator, however, realistically cause the affected instruments to wind down and slowly tumble. Demonstrating such a failure makes a lasting impression on students, usually because they fail to detect the problem and end up in an unusual attitude or crash. After that exercise, wary pilots keep the vacuum gauge and other indicators in their instrument scans.

For more information, see the topic "Setting up Failures" in the Learning Center.

The system failures in Flight Simulator are especially useful when controlled via the Flight Instructor Screen (in Flight Simulator 2004) and the Shared Aircraft feature in Flight Simulator X, described in Chapter 5, "Advanced 'Training Features' in Flight Simulator."

customizing Flight Simulator

Like most computer programs, Flight Simulator is infinitely adjustable. Other than customizing joystick buttons and adjusting some display settings, it is generally not necessary to fiddle with Flight Simulator if you plan to use it as a training aid.

Discussing all the customization features available in Flight Simulator is a topic beyond the scope of this book, but here are a few options to consider:

- Aircraft realism (precision of the flight model, fuel burn, etc.).
- Auto-rudder (eliminates the need for rudder pedals).
- Control sensitivities and "null zone" (response to joystick/yoke inputs).
- Setting the display resolution and level of detail shown on the screen.
- Customizing controls and key commands.

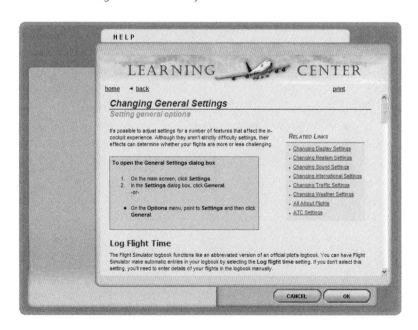

For more information, see the topic "Changing General Settings" and its related topics in the Learning Center, such as "Changing Realism Settings." You may also want to visit the official *Microsoft Flight Simulator* website, where you can find many articles about customizing and adjusting Flight Simulator.

five

advanced
"training features"
in Flight Simulator

checklist

☐ IFR training panels

☐ flight instructor
  utility

☐ shared aircraft

Microsoft Flight Simulator was developed primarily to create an environment that makes the illusion of flight as realistic and compelling as possible on currently available, general-purpose PCs. As the previous chapters explain, however, many of the "entertaining" features in Flight Simulator also can support learning and practicing real-world flying skills efficiently and effectively.

This chapter describes additional features that enhance the utility of Flight Simulator as a training aid, specifically:

- IFR Training Panels
- Flight Instructor Utility (*Microsoft Flight Simulator 2004*)
- Shared Aircraft (*Microsoft Flight Simulator X*)

IFR training panels

Flight Simulator 2004 includes versions of several aircraft with "IFR training panels." These panels—designed to help pilots who want to learn or practice IFR skills—combine the flight instruments and avionics in the main Flight Simulator window so that the important controls and instruments always remain in view. The space allocated to scenery is small—but because you practice IFR procedures in the clouds, the outside view isn't important. In effect, the large IFR panels mimic the view provided in many purpose-built flight training devices and PCATDs.

IFR training panel in the C172S

In Flight Simulator 2004, IFR training panels are available for the Cessna 172SP Skyhawk, the Cessna Skylane 182S, and Mooney M20M Bravo. To fly an aircraft with an IFR training panel, in the **Select Aircraft Dialog** box, select the "IFR panel" variation of the basic aircraft model that you want to fly.

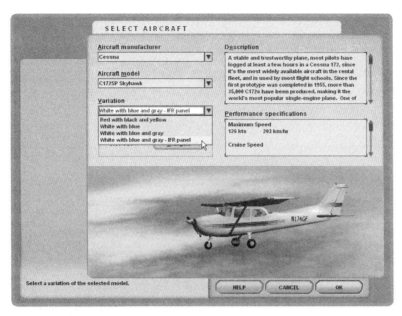

In Flight Simulator 2004, use the Select Aircraft dialog box to choose an aircraft with an IFR panel.

In Flight Simulator X, an IFR training panel is available for the Cessna 172SP Skyhawk. The procedure for loading the IFR panel is slightly different. First, select the normal cockpit view, and then press the **W** key to cycle through the series of panel views until you see the IFR panel or click the IFR panel icon on the instrument panel.

In Flight Simulator X, you can click a control panel icon to select the IFR panel in the Cessna 172.

advanced "training features" in flight simulator

flight instructor's utility

Microsoft Flight Simulator 2004 includes a training tool called the Flight Instructor Utility, which uses the simulator's built-in multiplayer feature to allow an instructor on one computer to monitor a student flying on a second PC. The student and instructor computers can be connected on a local network or via the Internet. In other words, they can be anywhere in the world.

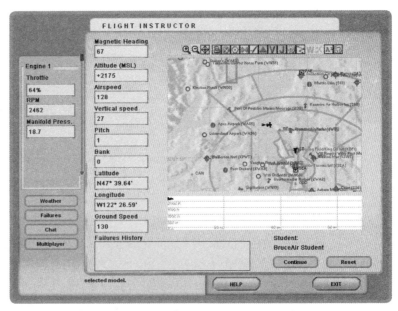

Main screen, flight instructor utility

While the student flies, the instructor sees the student's airplane overlaid on top of a moving map. The instructor can use the controls in the flight instructor utility to challenge students by changing the weather or by failing engines, instruments, and systems without interrupting the flight. During a flight, the instructor and student can chat by typing messages in pop-up text boxes.

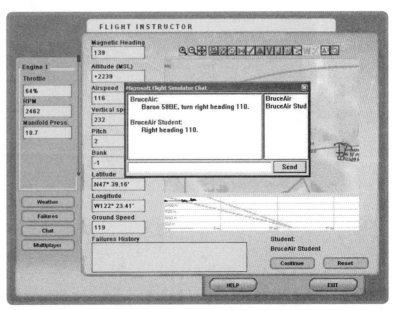

Chat window in FS2004 available in the flight instructor utility

using the flight instructor utility

To play the role of instructor, make sure that Flight Simulator isn't running on your PC, and then click the icon for Flight Instructor utility on the Windows desktop. Use the controls on the **Multiplayer Host Server** screen to create a multiplayer session and give it name. You can specify a password if you want to ensure that only a student with the proper information can join the session.

The student then starts Flight Simulator normally on a second PC and uses the Multiplayer feature in Flight Simulator to join the session hosted by the instructor.

Use the Multiplayer command on the Flights menu to join a multiplayer session.

Use the controls in the Multiplayer dialog box to manage a multiplayer session.

advanced "training features" in flight simulator

After the student joins the session, Flight Simulator runs normally on the trainee's PC. The student can load a Flight, create a Flight, select an aircraft, and so on. The instructor monitors the progress of the student's aircraft on the moving map, issues instructions with the chat feature (if the two computers aren't in the same room), and, in the tradition of instructors everywhere, "breaks things" by simulating failures.

Either the instructor or the student can leave a session at any time. If you are the instructor, click the **Exit** button in the **Flight Instructor** dialog box. If you are the student, on the **Flights** menu, click the **Multiplayer** command and then click **Disconnect**.

For more information, see the topic "Using the Flight Instructor Utility" in the Learning Center.

shared aircraft (*Microsoft Flight Simulator X*)

The multiplayer function in *Microsoft Flight Simulator X* introduces the "Shared Aircraft" feature, a concept that adds fun to virtual flying and enhances the roles that Flight Simulator can play as a training aid. It replaces the flight instructor utility from Flight Simulator 2004.

The Shared Aircraft feature allows two virtual aviators (*Microsoft Flight Simulator X* must be installed on both PCs) to fly the same aircraft simultaneously, as a crew or as a student and instructor. One pilot acts as host and chooses a starting location, aircraft, weather, and other initial conditions for the flight. The other pilot then joins the flight.

The Shared Aircraft feature synchronizes aircraft type, position, speed, heading, altitude, and other details. Each pilot, however, can select a separate view. For example, one pilot can fly in normal cockpit view while the other uses the virtual (3D) cockpit. Both pilots see the same scenery and weather from their respective cockpits. Each pilot can operate the avionics and most aircraft systems, but only one person can manipulate the flight and engine controls. Just as in a real aircraft, however, either pilot can pass primary control to the other.

While participating in a Shared Aircraft session, the two pilots can talk to one another as if using a voice-activated intercom (both computers must be equipped with microphones and speakers or headsets).

Because this feature is based on the multiplayer functions in Flight Simulator, two Flight Simulator users can connect over a local network or via the Internet and share a cockpit even if they're physically separated by thousands of miles.

The Shared Aircraft feature brings a new level of sophistication and utility to those who want to use Flight Simulator as a training aid. For example, now an instructor and student can fly a truly interactive virtual lesson without meeting at a flight school. Instead of just observing the student's flight, the instructor can help the student manage the aircraft and, when necessary, take over control to demonstrate tasks and procedures. An instructor who hosts a Shared Aircraft session can also set up system failures, weather, and other challenges for a student who joins the flight.

Equally important, with the Shared Aircraft feature, pilots can practice Crew Resource Management (CRM)—the effective use of all available resources and systematic sharing of responsibilities between the Pilot Flying (PF) and Pilot Not Flying (PNF). Although many people think of CRM as applying only to professional flight crews operating large aircraft, the concept is also important in other operations, such as flight instruction—in other words, whenever there is more than one person in the aircraft with access to the controls, avionics, and systems. With the Shared Aircraft feature, students, instructors, and aspiring professional pilots can practice CRM procedures on the Flight Simulator platform with all the advantages it offers over traditional FTDs and simulators.

With the Shared Cockpit feature in Flight Simulator X, two pilots can operate as crew, as if in a "real" simulator. (Alteon photo)

The Shared Cockpit feature in Flight Simulator X also allows an instructor to ride along with a student as if in a real airplane or FTD.

For more information about CRM, see FAA Advisory Circular 120-51E "Crew Resource Management Training." That AC includes many references to additional CRM resources.

advanced
"training
features"
in flight
simulator

six

differences between Flight Simulator 2004 and Flight Simulator X

checklist

☐ loading and using flights

☐ the view system

☐ using the mouse

☐ shared aircraft

☐ the G1000 cockpit

Microsoft Flight Simulator X was still in development as this book went to press. Some features, menus, dialog boxes, and other elements of the new version were still in flux, and late changes may affect the way the Practice Flights work in the new version. For the latest information about *Microsoft Flight Simulator X*, visit the official *Microsoft Flight Simulator* website at **www.fsinsider.com**. Late-breaking information about the Practice Flights and how they work with Flight Simulator X is available on the website for this book and in the "Latest Information" document on the CD included with this book.

loading and using flights

Flight Simulator X introduces "Missions," an updated version of the "Adventures" feature in earlier versions of Flight Simulator. Missions are Flights that include interactive elements, goals, and scoring. As in previous versions of Flight Simulator, however, Flights just set initial conditions.

The Practice Flights are available in sets for Flight Simulator 2004 and Flight Simulator X to ensure that Practice Flights initialize properly in Flight Simulator X. Otherwise, the two sets of Practice Flights are identical.

To load a Practice Flight in Flight Simulator X, use the **Free Flight** command on the main Flight Simulator screen and then click the **Load Flight** button. Or, on the **Flights** menu in Flight Simulator, choose the **Load Flight** command.

For more information about using Flights in Flight Simulator X, see the Learning Center.

In Flight Simulator X, use the Load button in the Free Flight dialog box to start the Practice Flights.

view system

Flight Simulator X includes many enhancements to the view system, including view "categories" and a "landing panel." The new view system also changes the way you switch to the IFR training panel in the Cessna 172SP Skyhawk. If you are flying Flight Simulator X, instead of choosing the **Select Aircraft** command on the **Aircraft** menu to pick the IFR Training Panel variant of the C172SP, just click the IFR Panel icon on the control panel. For more information, see the topic "Using Views and Windows" in the Learning Center.

The IFR panel for the Cessna 172 in Flight Simulator X.

Flight Simulator X also uses "alpha blending" to create see-through menus and instrument panels. For example, when you display the multifunction display (moving map) in the G1000 version of the Cessna 172SP Skyhawk, you can see the controls and map on the MFD, and you can also see the scenery outside the airplane.

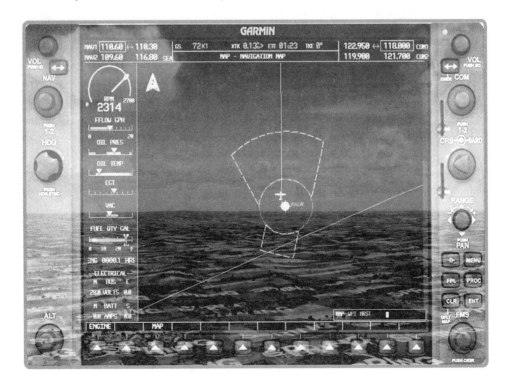

using the mouse

Flight Simulator X re-introduces the "mouse as yoke" feature from early versions of *Microsoft Flight Simulator*. When you select this option, you can use the mouse as if it were a joystick or yoke. I do not recommend this method of aircraft control—it can be extremely awkward and imprecise. Flying with the "mouse as yoke" function may also interfere with using the mouse to operate some cockpit controls as described in Chapter 4, "Flight Simulator Essentials." Fly the airplane with a joystick or yoke, and, as recommended in Chapter 4, use the mouse as an extension of your hand to operate cockpit controls.

For more information about using the mouse in Flight Simulator X, see the topic "Using the Mouse" in the Learning Center.

shared aircraft

The new Shared Aircraft feature, discussed in Chapter 5, "Advanced 'Training Features' in Flight Simulator," replaces the Flight Instructor utility in Flight Simulator 2004. It greatly expands the utility of Flight Simulator as a training aid. To take full advantage of its capabilities, you should review the topics about Shared Aircraft in the Learning Center.

the Garmin G1000 glass cockpit

If you have the deluxe version of Flight Simulator X, you can display a Garmin G1000 integrated avionics "glass cockpit" while flying the Cessna 172SP Skyhawk and the Beechcraft BE58 Baron used in the Practice Flights associated with this book.

To use the G1000 panels, load a Practice Flight and change to the G1000 variant of either the Cessna 172SP Skyhawk or Beechcraft BE58 Baron.

Flight Simulator X emulates the basic functions of the G1000 system, but like the simulations of the Garmin GPS units available in previous versions of Flight Simulator, the virtual system does not include all of the features of the real avionics.

The most important limitation of the simulation is that it can display only one of the main G1000 screens (Primary Flight Display or Multifunction Display) at any time. In a real aircraft, each display measures 10.4 inches diagonally. However, a typical computer monitor can't usefully and simultaneously display the PFD and MFD, outside view, and other cockpit controls. Just as with the Garmin G1000 Trainer (available for purchase from the Garmin website), you must switch between the displays as you fly. If you have a multi-monitor setup, however, you can take advantage of the advanced display features in Flight Simulator to show more than one display simultaneously. For more information about using more than one monitor with Flight Simulator, see the Learning Center and the official *Microsoft Flight Simulator* website.

Despite the limitations of the G1000 emulation, Flight Simulator X may prove especially valuable as training aids for pilots who are learning to fly the Garmin system. The Garmin G1000 Trainer, a PC-based tool that emulates most of the functions of the real system, has one major limitation: it does not include a flight model. For example, applying a left-aileron control input does not turn the airplane. It just makes the attitude indicator on the PFD show a bank. Similarly, applying an elevator input changes the pitch attitude and starts a climb or descent but does not affect airspeed. Adjusting the throttle also does not directly affect the engine instruments. In other words, you can't use the Garmin G1000 Trainer to hand-fly an instrument approach or fly other procedures. Using *Microsoft Flight Simulator* in conjunction with the Garmin G1000 Trainer, however, gives pilots the opportunity to learn the details of the system and see its basic features in action during a flight.

To learn more about the G1000 cockpit in Flight Simulator, see the topics about the system in the Learning Center. You can also download the complete set of G1000 operating handbooks from the Garmin website.

seven

about the practice flights

checklist

- ☐ what is a flight?
- ☐ how to use the practice flights
- ☐ file naming conventions
- ☐ starting in the air
- ☐ aircraft in practice flights
- ☐ flying your airplane
- ☐ locations
- ☐ ATC

It may seem from Chapter 2, "Using Flight Simulator as a Training Aid" and Chapter 3, "Best Practices for Using Flight Simulator," that Flight Simulator is a wonderful tool for everything *except* hands-on-the-controls simulated flying. Up to now, the many references to the Practice Flights associated with this book have only hinted at what the Practice Flights are and how you can use them as an instructor, pilot, or student. This chapter answers questions about the Flights feature in Flight Simulator, introduces the Practice Flights created for this book, and offers additional guidance about how best to use the Practice Flights to complement flight training.

what is a flight?

A *Flight* in *Microsoft Flight Simulator* is a set of initial conditions, including:

- An aircraft type and its configuration (position of landing gear and flaps, engine state, systems status, and so forth).
- The aircraft's position (and, for most of the Practice Flights, which begin in the air, the aircraft's speed, altitude, and heading).
- Avionics configuration (initial frequencies set in the communications and navigation radios, courses selected on navigation controls and indicators).
- Autopilot configuration (unless indicated otherwise, the Practice Flights begin with the autopilot ON with ALT and HDG modes selected).
- Weather (wind, clouds, visibility, precipitation), date, and time of day.
- Flight plan waypoints loaded into the GPS, when appropriate.
- The simulation state. (Unless indicated otherwise, the Practice Flights begin with the simulation paused. To start flying, press **P**.)

The Beechcraft BE58 Baron cockpit at the start of a typical Practice Flight

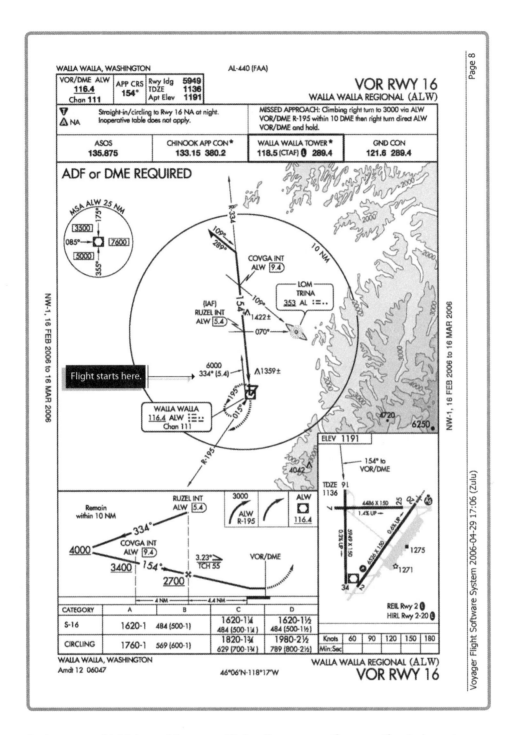

Saving a set of initial conditions as a Flight allows you to fly a specific airplane at a particular location, in a defined environment with the airplane in an appropriate configuration. This way, you do not have to set up every element manually each time you want to fly. For example, by loading a Flight, you can repeatedly practice flying an approach to your hometown airport without selecting an aircraft, repositioning the aircraft at the airport, adjusting the weather, taking off, and flying to the point where you want to start the approach. Set up everything once, save that situation as a Flight, and then, with just a couple clicks of the mouse, you can start flying from that position under those specific conditions whenever you like. In other words, each Flight is a template.

about the
practice
flights

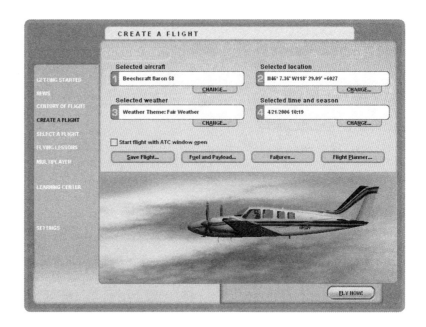

The Practice Flights associated with this book differ from the Flights supplied with Flight Simulator (such as the historical flights in Microsoft Flight Simulator 2004) only in purpose. The goal of each Practice Flight is to make it easy to learn about and practice a *specific skill or task*, such as basic attitude instrument flying, VOR navigation, entering and maintaining a holding pattern, or flying a particular type of instrument approach procedure. Practice Flights provide a starting point for a wide range of situations useful in training for VFR and IFR flying. In fact, the Practice Flights are designed to complement the typical training syllabi used in formal flight training.

more information about practice flights and flights

Practice Flights are based on the Flights feature in *Microsoft Flight Simulator*. None involves special programming. Flight Simulator X Flights can include events triggered by elapsed time, system failures, and other factors, including dialog and visual feedback. However, to stay focused on the goals of this book, and to make the Flights useful for many purposes, the Practice Flights for Flight Simulator X use the Flights feature only to set initial conditions, just as Flights work in Flight Simulator 2004.

To learn more about using the Flights feature in Flight Simulator 2004, see the "All About Flights" topic in the Learning Center.

collections of flights

The BruceAir Practice Flights that accompany this book are organized into two categories, VFR Practice Flights and IFR Practice Flights.

The VFR Practice Flights focus on skills and tasks associated with a syllabus for a private pilot certificate. The IFR Practice Flights enhance training for the instrument rating. Some Practice Flights may be useful for more than one training situation. For example, the Basic Attitude Instrument Flying and VOR Navigation Practice Flights cover skills and procedures that both VFR students and instrument students must learn.

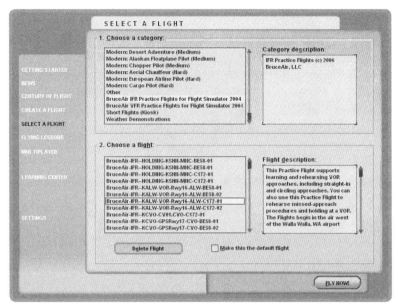

The Practice Flights described in this book are all labeled "BruceAir-VFR" or "BruceAir-IFR"

note

The complete, up-to-date list of all the Practice Flights associated with this book is available in Adobe Reader format (.pdf) on the CD included with this book and on the website for this book. You can find general descriptions and other information about the VFR and IFR categories of Practice Flights later in this book.

practice flight descriptions

Each Practice Flight includes a general description to help you quickly choose the most appropriate Flight or Flight for a specific task.

The description appears in the **Select a Flight** dialog box (Flight Simulator 2004) or **Flights** dialog box (Flight Simulator X), and it contains the following important information about the Practice Flight:

- Objective
- Type of aircraft
- Starting location (usually relative to a navaid or airport)
- Useful information and instructions specific to that Practice Flight

For example, here is a description for a Practice Flight about VOR navigation:

This Practice Flight supports learning and rehearsing basic VOR tracking skills. It begins in the air in day-VMC west of Walla Walla, WA (KALW) on V520 between the Walla Walla (ALW) and Pasco (PSC) VORs. You are at 4,500 feet in the Beechcraft BE58 Baron with the default instrument panel. Practice intercepting and tracking inbound and outbound on radials from VORs in the area and identifying intersections. The autopilot is ON in HDG and ALT modes, and the simulation is paused. To start flying, press the "P" key.

about the
practice
flights

preflight briefings

A detailed preflight briefing is provided in Adobe Reader format (.pdf) for each group of Practice Flights. The briefings are available on the CD that accompanies this book. You can print the briefings for review while using a Practice Flight. The briefing repeats the basic information from the description, and it offers other details to help you prepare for and benefit from the Practice Flight, including:

- Aircraft
- Location
- Initial conditions
- Recommended background reading

You can find examples of the briefings for VFR and IFR Practice Flight in Chapter 11, "Briefings for VFR Practice Flights," and Chapter 13, "Briefings for IFR Practice Flights."

installing the practice flights

To use the Practice Flights that accompany this book, just copy them from the CD to the appropriate folder on your PC. The specific procedure to follow depends on whether you use Flight Simulator 2004 or Flight Simulator X and whether you are running Windows XP or Windows Vista. The following sections provide general steps to follow for each situation. For the latest information about how to set up the Practice Flights, see "Installing the Practice Flights" (InstallPracticeFlights.pdf) on the CD and website for this book.

how the practice flights are organized

The Practice Flights are collected into two main folders, one for Flight Simulator 2004 users, and the other for people who fly Flight Simulator X. Each set of Practice Flights is further divided into two sub-folders, one for VFR Flights and another for IFR Flights.

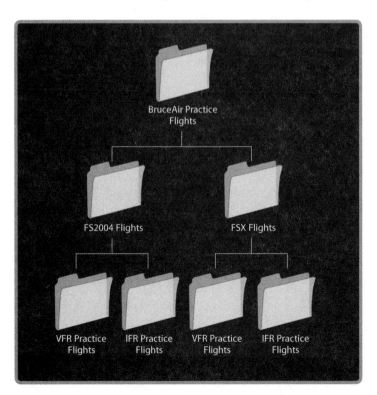

where to copy the practice flights

I recommend copying the folders that contain the BruceAir Practice Flights to the main Flight Simulator Flights folder on your hard drive. The default location of the Flights folder for Flight Simulator 2004 and Windows XP is:

C:\Program Files\Microsoft Games\Flight Simulator 9\Flights

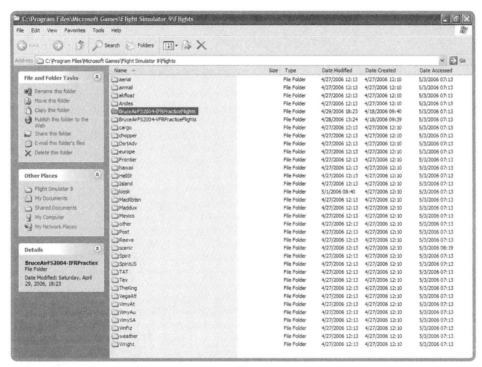

In Flight Simulator 2004, you can copy the folders for the BruceAir VFR and IFR Practice Flights into the main Flights folder.

To see the **Flights** folder and copy files to that folder, you must have administrator permissions on the computer and you must select the option to display system folders in Window Explorer.

To display the contents of system folders, open **Folder Options** in the Control Panel:

- Click **Start**, and then click **Control Panel**.
- Click **Appearance and Themes**, and then click **Folder Options**.
- On the **View** tab, select the **Display the contents of system folders** check box.

You can also copy all of the Practice Flights (not the folders for the IFR and VFR Practice Flights) to the **Flight Simulator Files** folder in your **My Documents** folder: C:\Documents and Settings\<Your Computer User Name>\My Documents\Flight Simulator Files. This procedure is the easiest and works regardless of which version of Flight Simulator or Windows you are using.

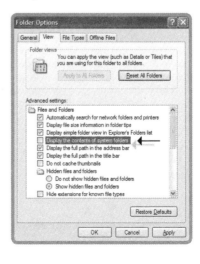

If you copy the BruceAir Practice Flights into the **Flight Simulator Files** folder, the Practice Flights won't appear in the main list of flights in the **Select Flights** dialog box and they won't be organized into VFR and IFR categories. To see the BruceAir Practice Flights, first select **My Saved Flights**. All of the Practice Flights will then appear in the list.

If you installed Flight Simulator to a different folder, copy the sets of Practice Flights to the appropriate location on your hard drive.

After you copy the Practice Flight folders and start Flight Simulator, you will see the Practice Flights in the **Select Flight** dialog box.

starting a practice flight

Starting a Practice Flight is much like opening any type of file on a computer. In Flight Simulator 2004, you can select a Flight from the opening screen or by using the Flights menu. (You can also launch a Practice Flight by double-clicking the appropriate .flt file in a folder list.)

Either method opens the **Select a Flight** dialog box, where you can review the list of Flights available on your computer and select a specific Flight to start.

In Flight Simulator X, you can select a Flight from the opening screen or by using the Flights menu.

Either method opens the **Flights** dialog box, where you can review the list of Flights available on your computer and select a specific Practice Flight to start.

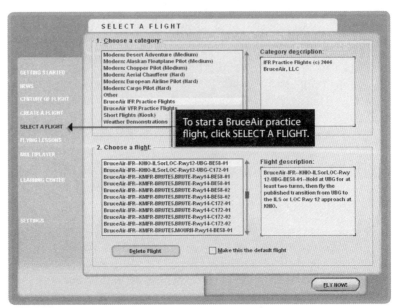

title and file naming conventions for practice flights

The titles and file names for the Practice Flights associated with this book follow a consistent pattern to make them easy to identify in the **Select a Flight** or **Flights** dialog box.

Each name begins with "BruceAir" and an abbreviation indicating whether it is a VFR or IFR Practice Flight. The identifier for the nearest airport or navaid follows. If the goal of a Practice Flight is to learn about an instrument procedure, the procedure title and transition come next. Finally, the file name includes the type of aircraft and an index number to distinguish among Practice Flights that share the same purpose, location, and aircraft, but differ in time of day and weather. The file name (as it appears in the listing of a folder's contents) includes the .flt extension.

For example, **BruceAir--VFR-Basic_VOR-KOLM-C172S-01.flt** indicates a VFR Practice Flight for practicing basic VOR skills. It begins near the Olympia, WA airport in the Cessna 172SP in day-VMC with light wind.

The variation **BruceAir--VFR-Basic_VOR-KOLM-C172S-02.flt** begins with the same conditions, except with winds aloft so that you can practice compensating for drift.

A typical IFR Practice Flight has a file name such as **BruceAir--IFR-ILS-KSFF-ILS-Rwy21R-AZTEM–BE58-01.flt**. This name indicates an IFR Practice Flight for practicing an ILS approach. The flight begins near the Felts Field airport at Spokane, WA with the aircraft set up to fly the ILS Rwy 21R approach via the transition from the AZTEM fix. The aircraft is the Beechcraft BE58Baron, and this version of the Practice Flight begins in day-VMC with no wind.

Such details are included in the Practice Flight file names to make each Practice Flight easy to identify and to help you quickly locate information about the airports and procedures in airport and facility guides.

variations on a theme

As the naming conventions imply, the basic situation captured in one Practice Flight forms the foundation for several related Flights. For example, several Practice Flights in the Holding Procedures category start at different locations near the same VOR so that you can easily experience a variety of hold entries and fly published as well as ad-hoc holding patterns. Some of the IFR Practice Flights begin in VMC to help you associate instrument indications with the view out the window. Others simulate IMC to provide a realistic IFR training environment.

flight, weather, and flight plan files

The basic information about a Flight is stored in a file with the extension .flt. Information about the weather associated with that Flight is collected in a file with the identical name and the extension .wx. If a Flight is linked to a flight plan (a route from departure to destination), the flight plan file uses the name of the Flight file but adds the extension .pln.

If you copy, move, or share Flight Simulator Flights, make sure you copy the weather file and, if appropriate, the flight plan file associated with each Flight.

Here is an example of a typical IFR Practice Flight and its associated weather file:

BruceAir--IFR-ILS-KSFF-ILS-Rwy21R-AZTEM–BE58-01.flt

BruceAir--IFR-ILS-KSFF-ILS-Rwy21R-AZTEM–BE58-01.wx

starting in the air

Most of the Practice Flights begin in the air, with the autopilot ON and with the HDG and ALT modes selected. As explained earlier in the book, this arrangement assures that each Practice Flight starts in a stable condition and allows you to concentrate on procedures and not just hand-flying the airplane. Starting with the autopilot on also makes it easier for instructors to use Practice Flights as interactive demonstration tools.

To turn the autopilot on or off:

Click the AP button on the autopilot.

-or-

Press the Z key

-or-

Click the AP annunciator on the instrument panel.

You can learn more about the autopilot in the Flight Simulator Learning Center and in "About the Autopilot," an article on the official Microsoft Flight Simulator website.

aircraft used in the practice flights

All of the Practice Flights use default aircraft provided with *Microsoft Flight Simulator 2004* or *Microsoft Flight Simulator X*, in particular, the Cessna 172SP and the Beechcraft BE58 Baron.

These aircraft represent basic types of aircraft—a single-engine trainer/personal airplane and a high-performance piston twin—common in the general aviation fleet. Their performance characteristics, instrument panels, avionics, and other controls also correspond well to those type airplanes. Focusing on just two aircraft keeps the number of variations for each Practice Flight reasonable. As detailed in Chapter 2, this still supports the most important goal of using Flight Simulator to enhance formal flight training.

Cessna 172P

Note that the Baron is included not just because it is a twin. Its performance and complexity support general learning skills associated with such airplanes, including retractable gear and a constant speed propeller. In other words, flying the Baron with both engines running (as is the case in all the Practice Flights) offers an operating environment comparable to a Bonanza and other high-performance aircraft, single or twin. The Baron instrument panel also includes an HSI and RMI—useful for practicing advanced IFR procedures.

The Cessna 172SP and the Baron are also available with the G1000 glass cockpits in *Microsoft Flight Simulator X*, and in both versions of Flight Simulator, the Cessna includes a variation with an IFR training panel. Both factors support using these aircraft as platforms for the Practice Flights.

Beechcraft BE58 Baron

flying your favorite airplane

Because it is impossible to anticipate which aircraft you may have added to your Flight Simulator hangar, I've based all the Flights on aircraft in the default fleet supplied with Flight Simulator.

Nothing, however, prevents you from flying your favorite aircraft or switching among other default Flight Simulator aircraft when you use the Practice Flights. Just load a Practice Flight, and before you start flying, open the **Aircraft** menu and click **Select Aircraft**. Note that if you choose another aircraft, you may have to retract or extend the landing gear, adjust power settings or make other changes to match the initial conditions specified when I created the Flight.

If you want to save a Practice Flight but preserve your choice of a favorite aircraft, load the Practice Flight, switch to another aircraft, adjust its configuration as necessary, and then save the Flight with a new name. The Flight you save is stored in the **My Saved Flights** folder on your hard drive. For more information about saving Flights, see the topic "All about Flights" in the Learning Center.

location, location, location

Most of the Practice Flights begin in the Pacific Northwest. That's where I do most of my flying and instructing, but it's also an excellent region in which to learn about flying—especially IFR flying (and not just because it rains a lot). The terrain, complex airspace, and other factors in the Pacific Northwest present a representative sample of VFR situations and IFR procedures that offer excellent training opportunities, regardless of where you fly.

However, you can use Flight Simulator to practice specific procedures in the areas where you fly, to preview specific cross-country trips, and to explore challenging situations. To learn more about creating your own Flights, see Chapter 14, "Creating Your Own Practice Flights," and the Learning Center.

use of ATC

Most of the Practice Flights do not use the ATC feature in *Microsoft Flight Simulator* because ATC is designed to work primarily on complete trips associated with a flight plan. The goal of these Practice Flights is to practice specific tasks in discrete, short sessions. In most cases, adding ATC to the mix unnecessarily complicates matters. Where appropriate, the introduction to each Practice Flight includes an appropriate clearance for you to follow. (To learn more about how to copy and read an ATC clearance in standard shorthand, see Appendix 1, "Clearance Shorthand," in the *Instrument Flying Handbook*.)

For added verisimilitude, I suggest making standard reports to ATC as you fly (e.g., when entering a hold, contacting the control tower at the outer marker during and ILS, etc.). After all, in the privacy of your home, who is going to notice that you're talking to yourself?

You can learn more about how the ATC system in Flight Simulator works in the Learning Center. Check out the entire ATC section and especially "IFR Flight and ATC" and "Real-World ATC Differences."

evaluating your flight

The Practice Flights are not lessons, and they do not include a scoring system. As noted above, they set initial conditions, making it easy for you practice specific procedures and skills.

You can, however, use the Flight Analysis feature to review your performance on many flights, whether you fly on your own or with an instructor. For more information, see Chapter 4, "Flight Simulator Essentials," and the "Flight Analysis" topic in the Learning Center.

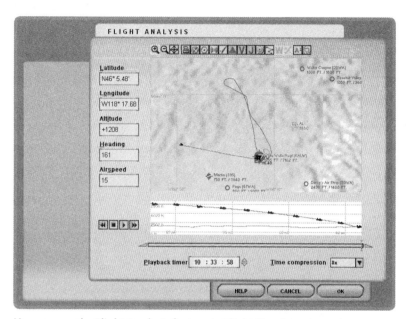

You can use the Flight Analysis feature in Flight Simulator to evaluate your performance.

eight

flying the aircraft used in the practice flights

checklist

☐ about cockpits

☐ flying "the numbers"

☐ flying the Skyhawk

☐ flying the Baron

The previous chapter explained why the Practice Flights associated with this book use the Cessna 172SP and the Beechcraft BE58 Baron, both which are part of the standard Flight Simulator fleet. This chapter describes each aircraft in more detail and provides specific operating tips to help you fly them smoothly and precisely.

You can find additional information about these airplanes, including checklists and flying hints, in the "Aircraft Information" section of the Learning Center.

general information about Flight Simulator cockpits

The aircraft supplied with Flight Simulator share a few key adaptations useful for virtual flying. For example, you won't find a row of control panel icons on a real airplane instrument panel.

Clicking these icons gives you a quick way to display (and hide) additional information and features—such as Kneeboard, GPS window, and avionics stack. The location and number of icons varies among aircraft, but the Skyhawk and the Baron used in the Practice Flights include the following icons:

- Kneeboard
- ATC
- Map
- Avionics
- GPS
- Magnetic compass
- Engine Controls (Baron)

For more information about the control panel icons, see the topic "Using the Mouse" in the Learning Center.

In addition, remember that you can operate most of the controls and switches on the instrument panel with the mouse, reducing or eliminating the need to have a keyboard handy in flight. For more information, see Chapter 4, "Flight Simulator Essentials."

flying "the numbers"

Chapter 3, "Best Practices for Using Flight Simulator," explained the concept of using "the numbers" to stay ahead of the airplane and reduce workload, especially when flying IFR.

The tables provided for the Skyhawk and Baron later in this chapter offer guidance to help you set up common configurations such as climbs, cruise, descent, and approach. The numbers in the tables are not hard-and-fast rules. For example, in an airplane like the Skyhawk, rate of climb decreases as you gain altitude. You won't be able to maintain a climb rate of 700 fpm all the way up to 10,000 ft. However, if you're at about 5,000 ft. or less, the numbers get you close to the mark. Use these numbers to get the airplane stabilized in a fundamental configuration. You can expand the table by experimenting with settings for other situations.

To use the tables, pick a situation, and then follow the "pitch-power-configuration-trim" sequence to stabilize the airplane in that condition:

- Use the control yoke or joystick to set the airplane's attitude so it matches the value in the "Attitude" column and the picture above the table.
- Set the throttle so that the reading on the tachometer or manifold pressure gauge matches the value in the "RPM or MP" row. When flying the Baron, use the throttle to set manifold pressure and use the propeller control to set the RPM.
- Use the flap and landing gear controls to configure the airplane as specified in the table.

- Allow the airplane to stabilize, and apply nose-up or nose-down elevator trim so that the airplane flies "hands-off."
- Check the airspeed indicator and vertical speed indicator, and then make small, smooth adjustments to the power and/or aircraft attitude to fine-tune the airplane's performance.

Chapter 3, "Best Practices for Using Flight Simulator," describes how you can also use the autopilot to establish different configurations.

fine-tuning and trim

After you establish a basic configuration using the numbers in the tables, make small adjustments to the pitch attitude and power in order to achieve the exact performance you want. "Small" means increments of no more than 100 RPM or 1 in. of MP and one or two degrees of pitch. The thickness of the wings on the attitude indicator's miniature airplane is a good guide for how much to change the pitch attitude each time you want to fine-tune the aircraft's performance.

This book is not the place to join the great "pitch v. power" debate about whether pitch or power is primary or secondary during different phases of flight. If you want to learn more about the concept, a good introduction is "Pitch and Power" in Chapter 3, "Basic Flight Maneuvers" of the *Airplane Flying Handbook* (FAA-H-8083-3A). The discussions of the fundamental flight maneuvers throughout that chapter and in Chapter 5, "Airplane Basic Flight Maneuvers," of the *Instrument Flying Handbook* (FAA-H-8083-15) also provide detailed, practical advice.

As a general rule, just remember that each time you adjust the pitch attitude you'll also need to change the power setting, and vice-versa (assuming you want to maintain airspeed). Change one thing at a time and give the airplane a few moments to stabilize after you adjust either power or pitch. Patience is key.

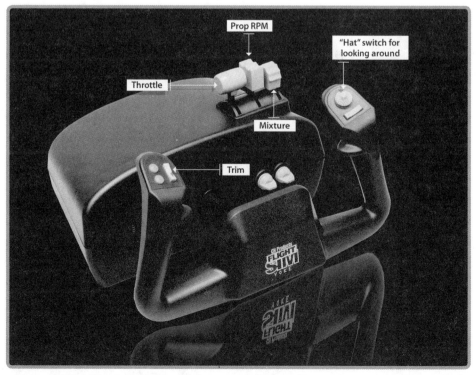

Most joysticks and yokes, like this Flight Sim Yoke from CH Products, include all the basic controls you need.

flying
the aircraft
used in the
practice
flights

Finally, use elevator trim to help you maintain precise control of the airplane. If, after setting pitch, power, and flaps/gear you let the airspeed stabilize and find that forward pressure is needed to maintain attitude, add nose down trim until you can relax pressure on the stick. If you must hold a little back pressure on the stick to maintain the pitch attitude, add nose-up trim until you can relax the pressure on the joystick. Remember that you can also use the mouse to point to the trim wheel on the instrument panel and then roll the mouse wheel forward or back to add up and down trim.

The following key points about the use of trim from the *Airplane Flying Handbook* (FAA-H-8083-3A) and the *Instrument Flying Handbook* (FAA-H-8083-15) are worth noting here:

> *Changes in attitude, power, or configuration will require a trim adjustment, in most cases. Using trim alone to establish a change in aircraft attitude invariably leads to erratic aircraft control. Smooth and precise attitude changes are best attained by a combination of control pressures and trim adjustments. Therefore, when used correctly, trim adjustment is an aid to smooth aircraft control.[1]*

The handbook adds that errors in using the trim often arise from:

- Faulty sequence in trim technique. "Trim should be used, not as a substitute for control with the wheel (stick) and rudders, but to relieve pressures already held to stabilize attitude. As you gain proficiency, you become familiar with trim settings, just as you do with power settings. With little conscious effort, you trim off pressures continually as they occur."
- Large trim inputs. "This practice induces control pressures that must be held until you retrim properly. Use trim frequently and in small amounts."[2]

flying the Cessna 172SP Skyhawk

1 *Instrument Flying Handbook* (FAA-H-8083-15), 5–14
2 *Instrument Flying Handbook* (FAA-H-8083-15), 5–14

cockpit orientation

The Cessna 172SP in Flight Simulator is available in several versions, each with a different instrument panel:

- The standard Flight Simulator cockpit includes a primary instrument panel, a pop-up avionics stack, and a large outside view above the instrument panel.
- The IFR Panel features a large instrument panel that includes all controls and instruments, including the avionics stack and magnetic compass, with a small outside view across the top of the cockpit window.
- The "glass cockpit" version simulates the Garmin G1000 integrated avionics system (available only in *Microsoft Flight Simulator X*).

The Cessna 172SP variations share the same engine, performance, and handling characteristics, so the operating tips described in this chapter apply to all three versions of the Skyhawk.

conventional cockpit

The standard instrument panel of the Cessna 172SP in Flight Simulator should be immediately familiar to anyone who has logged time in the venerable Skyhawk or a comparable aircraft.

The flight, engine, and systems instruments and basic controls are arranged conventionally, with only a few deviations (e.g., the location of the elevator trim wheel and fuel selector) from the layout in a real, late-model Skyhawk.

The avionics and GPS in the standard Skyhawk appear in pop-up windows, which you can display and hide by clicking icons on the control panel.

For more information about the standard cockpit layout, see the topic "Cockpit Basics" in the Learning Center.

Actual Cessna 172 instrument panel

flying
the aircraft
used in the
practice
flights

121

C172 instrument panel in Flight Simulator 2004 with the avionics stack displayed.

C172 instrument panel in Flight Simulator 2004 with the GPS displayed.

IFR training panel

The IFR Panel version of the Skyhawk features a large instrument panel that consolidates all controls and instruments (including the avionics stack and magnetic compass) in the main cockpit display. This arrangement simplifies IFR flights, because you don't have to display the avionics stack to tune radios and use the autopilot.

For more information about using the IFR training panels, see Chapter 5, "Advanced 'Training Features' in Flight Simulator."

C172 IFR training panel in Flight Simulator 2004.

Garmin G1000 glass cockpit

The Garmin G1000 version of the Skyhawk (available only in *Microsoft Flight Simulator X Deluxe Edition*) simulates the core features of the Garmin G1000 integrated avionics system. The limited space available on a conventional PC monitor means that you can show only one of the two main displays in the G1000 "glass cockpit" at any time. You must switch between the Primary Flight Display (PFD), which contains all of the flight instruments and important supplemental information, and the Multifunction Display (MFD), which features a large moving map and instruments that show the status of the engine and aircraft systems.

G1000 PFD in Flight Simulator X Deluxe Edition

flying
the aircraft
used in the
practice
flights

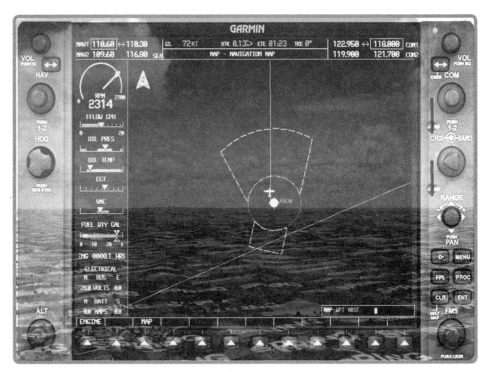

G1000 MFD in Flight Simulator X Deluxe Edition

Explaining the details of the G1000 system is beyond the scope of this book. For more information, see the G1000 topics in the Learning Center for *Microsoft Flight Simulator X*. You can also download the complete set of G1000 manuals in Adobe Reader format (.pdf) from the Garmin website.

"the numbers" for flying the Skyhawk

The following tables provide guidance to help you configure the Cessna 172SP for the following common situations:

- Initial Climb
- Cruise Climb
- Normal Cruise
- Low-Speed Cruise
- En Route Descent
- Level Flight in the Traffic Pattern
- Final Approach
- ILS Descent (Flaps)
- ILS Descent (No Flaps)

All of the configurations assume that the aircraft has full fuel tanks and is operating at near its maximum allowable gross weight.

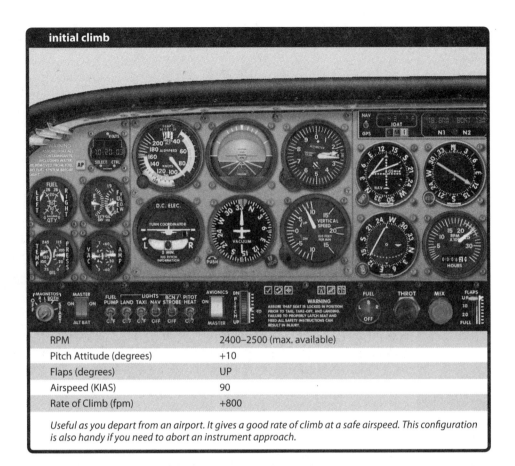

initial climb

RPM	2400–2500 (max. available)
Pitch Attitude (degrees)	+10
Flaps (degrees)	UP
Airspeed (KIAS)	90
Rate of Climb (fpm)	+800

Useful as you depart from an airport. It gives a good rate of climb at a safe airspeed. This configuration is also handy if you need to abort an instrument approach.

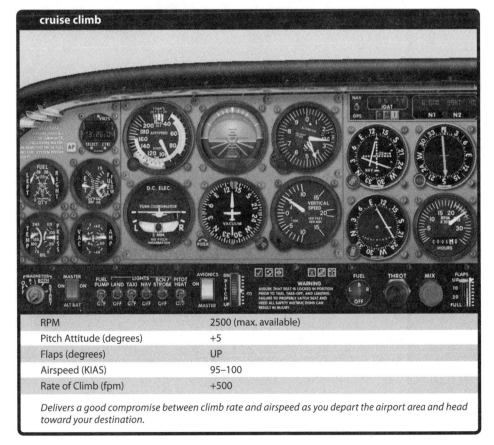

cruise climb

RPM	2500 (max. available)
Pitch Attitude (degrees)	+5
Flaps (degrees)	UP
Airspeed (KIAS)	95–100
Rate of Climb (fpm)	+500

Delivers a good compromise between climb rate and airspeed as you depart the airport area and head toward your destination.

flying
the aircraft
used in the
practice
flights

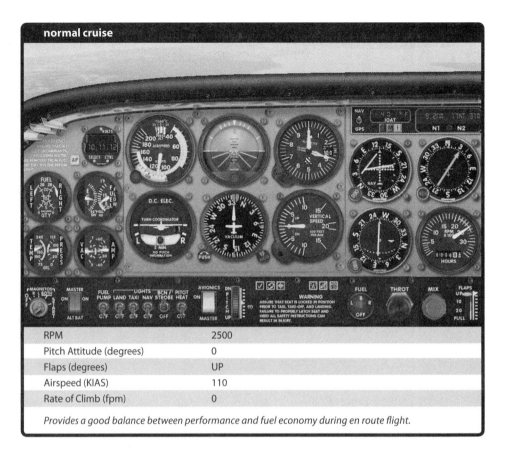

normal cruise

RPM	2500
Pitch Attitude (degrees)	0
Flaps (degrees)	UP
Airspeed (KIAS)	110
Rate of Climb (fpm)	0

Provides a good balance between performance and fuel economy during en route flight.

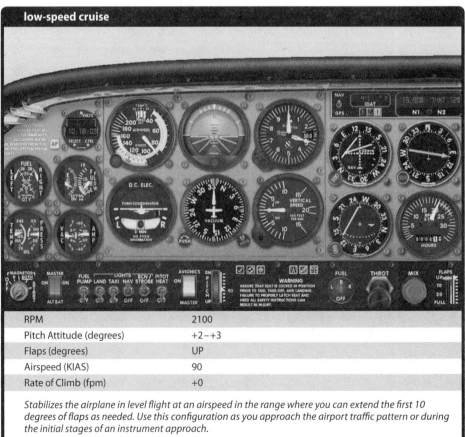

low-speed cruise

RPM	2100
Pitch Attitude (degrees)	+2 – +3
Flaps (degrees)	UP
Airspeed (KIAS)	90
Rate of Climb (fpm)	+0

Stabilizes the airplane in level flight at an airspeed in the range where you can extend the first 10 degrees of flaps as needed. Use this configuration as you approach the airport traffic pattern or during the initial stages of an instrument approach.

en route descent

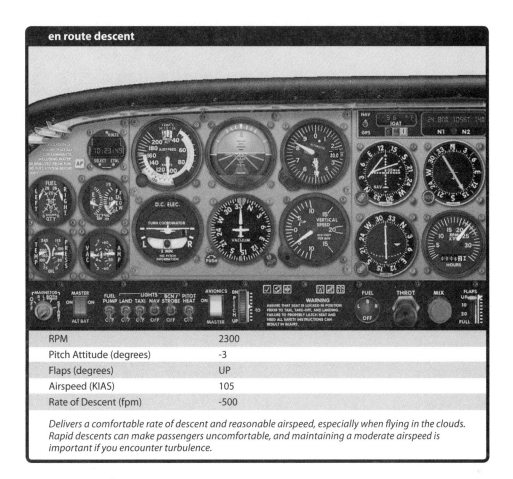

RPM	2300
Pitch Attitude (degrees)	-3
Flaps (degrees)	UP
Airspeed (KIAS)	105
Rate of Descent (fpm)	-500

Delivers a comfortable rate of descent and reasonable airspeed, especially when flying in the clouds. Rapid descents can make passengers uncomfortable, and maintaining a moderate airspeed is important if you encounter turbulence.

level flight in the traffic pattern

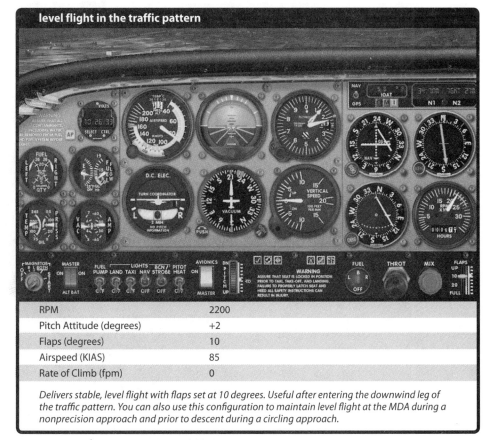

RPM	2200
Pitch Attitude (degrees)	+2
Flaps (degrees)	10
Airspeed (KIAS)	85
Rate of Climb (fpm)	0

Delivers stable, level flight with flaps set at 10 degrees. Useful after entering the downwind leg of the traffic pattern. You can also use this configuration to maintain level flight at the MDA during a nonprecision approach and prior to descent during a circling approach.

flying
the aircraft
used in the
practice
flights

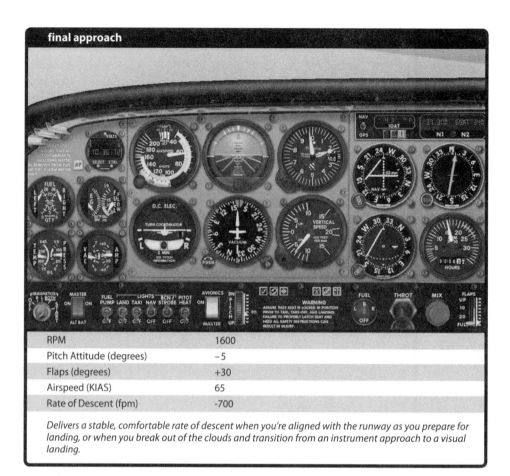

final approach

RPM	1600
Pitch Attitude (degrees)	−5
Flaps (degrees)	+30
Airspeed (KIAS)	65
Rate of Descent (fpm)	−700

Delivers a stable, comfortable rate of descent when you're aligned with the runway as you prepare for landing, or when you break out of the clouds and transition from an instrument approach to a visual landing.

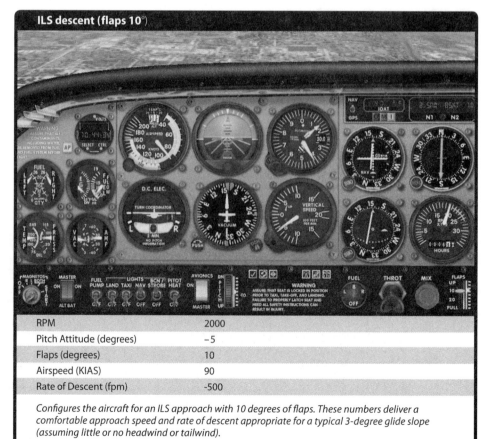

ILS descent (flaps 10°)

RPM	2000
Pitch Attitude (degrees)	−5
Flaps (degrees)	10
Airspeed (KIAS)	90
Rate of Descent (fpm)	−500

Configures the aircraft for an ILS approach with 10 degrees of flaps. These numbers deliver a comfortable approach speed and rate of descent appropriate for a typical 3-degree glide slope (assuming little or no headwind or tailwind).

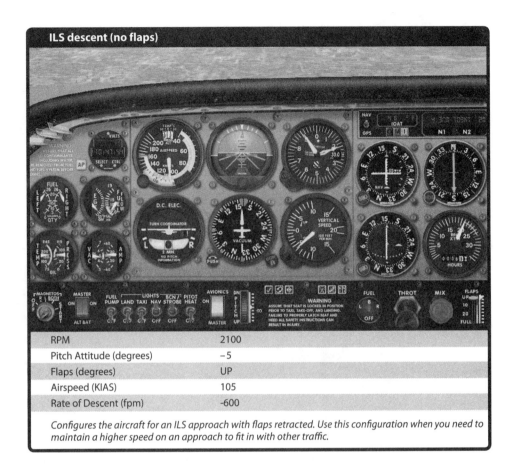

ILS descent (no flaps)

RPM	2100
Pitch Attitude (degrees)	–5
Flaps (degrees)	UP
Airspeed (KIAS)	105
Rate of Descent (fpm)	-600

Configures the aircraft for an ILS approach with flaps retracted. Use this configuration when you need to maintain a higher speed on an approach to fit in with other traffic.

flying the Beechcraft BE58 Baron

flying the aircraft used in the practice flights

cockpit orientation

The standard instrument panel for the Beechcraft BE58 Baron in Flight Simulator is typical of the arrangement used in high-performance, complex aircraft, and it is nearly identical to the layout used in the Beechcraft A36 Bonanza.

The flight and systems instruments and basic controls are arranged conventionally, with only a few deviations from the cockpit in a real, late-model Baron or Bonanza (e.g., the location of the trim wheels).

The Beechcraft BE58 instrument panel in Flight Simulator.

RMI and HSI from the BE58 panel

Note that the Baron instrument panel includes a Horizontal Situation Indicator (HSI) and a dual-needle Radio Magnetic Indicator (RMI). The HSI displays navigation information from the NAV 1 receiver or GPS. The RMI displays navigation information from the NAV 2 receiver and ADF receiver. Both instruments feature "slaved" compass cards that always display the aircraft's current heading. These instruments are typically found in high-performance single-engine and twin-engine aircraft. To learn more about them, see "Remote Indicating Compass" in Chapter 3, "Flight Instruments," and Chapter 5, "Navigation Systems" in the *Instrument Flying Handbook* (FAA-H-8083-15).

Like the Skyhawk, the Baron uses pop-up windows to display the avionics stack and GPS.

The engine and fuel controls for the Baron are also displayed in a pop-up window.

Baron panel with avionics stack displayed.

flying
the aircraft
used in the
practice
flights

Baron with GPS displayed

Baron panel with engine controls displayed

You can display and hide each of these pop-up windows by clicking icons on the control panel.

For more information about the standard cockpit layout, see "Cockpit Basics" in the Learning Center. For more information about the Throttle Quadrant window, see "Using Views and Windows" in the Learning Center.

setting power in the Baron

The engine controls on the Baron are not visible in the normal cockpit view, but you can use keyboard shortcuts to adjust the throttles, propeller RPM, and mixtures. To move the controls with the mouse, open the Throttle Quadrant pop-up window by clicking the airplane icon on the left side of the panel, and then drag the levers that you want to adjust. (Note that you cannot use the mouse wheel to move the throttles, prop levers, or mixture controls in the Baron. You must point to the lever(s) that you want to adjust and then drag them with the mouse.)

If you prefer not to display the engine controls, you can adjust the throttle and propeller RPM for both engines simultaneously by using the keyboard as described below.

adjusting the throttle: On the Baron, the throttles control manifold pressure, which is displayed in inches of mercury on the manifold pressure gauges. You can increase or decrease the manifold pressure with the throttle control on a joystick or yoke, by dragging the throttle with a mouse, or by pressing F3 to increase power and F2 to decrease power. Remember that as you climb into less dense air, the manifold pressure decreases by about one inch per thousand feet. To maintain a specific manifold pressure setting, you must increase the throttle as you climb. Note that above about 10,000 feet, however, you will no longer be able to maintain manifold pressure settings greater than approximately 21 inches. As you descend with a constant throttle setting, manifold pressure increases at the same rate—about one inch per thousand feet. To maintain a specific manifold pressure setting as you descend, you must gradually decrease the throttle.

setting propeller RPM: The prop levers set propeller speed, which is shown on the tachometers in hundreds of revolutions per minute (RPM). You can drag the propeller control with the mouse or press CTRL+F3 to increase RPM or CTRL+F2 to decrease RPM. Note that if you reduce manifold pressure below the bottom of the green arc on the manifold pressure gauge (i.e., to less than about 15 in.) and indicated airspeed decreases, the propeller governor cannot maintain the last RPM you set. As you continue to reduce the throttle, both the manifold pressure and RPM decrease.

For more information about operating an engine equipped with a constant-speed propeller, see "Basic Propeller Principles" in Chapter 5 "Aircraft Systems" of the *Pilot's Handbook of Aeronautical Knowledge* (FAA-H-8083-25) and "Constant-Speed Propeller" in Chapter 11, "Transition to Complex Airplanes" in *Airplane Flying Handbook* (FAA-H-8083-3A).

Garmin G1000 glass cockpit

The Garmin G1000 version of the Baron (available only in *Microsoft Flight Simulator X Deluxe Edition*) simulates the core features of the Garmin G1000 integrated avionics system. The limited space available on a conventional PC monitor means that you can show only one of the two displays in the G1000 "glass cockpit" at any time. You must switch between the Primary Flight Display (PFD), which contains all of the flight instruments and important supplemental information, and the Multifunction Display (MFD), which features a large moving map and instruments that show the status of the engines and aircraft systems.

G1000 PFD for the Baron in Flight Simulator X Deluxe Edition

flying
the aircraft
used in the
practice
flights

133

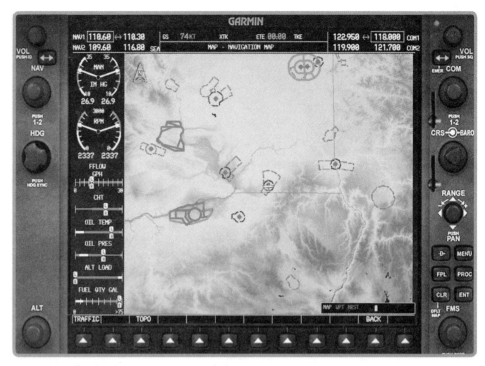

G1000 MFD for the Baron in Flight Simulator X Deluxe Edition

Explaining the details of the G1000 system is beyond the scope of this book. For more information, see the G1000 topics in the Learning Center for *Microsoft Flight Simulator X*. You can also download the complete set of G1000 manuals in PDF format from the Garmin website.

"the numbers" for flying the Beechcraft BE 58 Baron

The following tables provide guidance to help you configure the Baron for the following common situations:

- Initial Climb
- Cruise Climb
- Normal Cruise
- Low-Speed Cruise
- En Route Descent
- Expedited Descent
- Transition Speed—Level Flight
- Level Flight in the Traffic Pattern
- Short Final
- IFR Approach Intermediate Maneuvering
- ILS Descent

All of the configurations assume that the aircraft has full fuel tanks and is operating at near its maximum allowable gross weight.

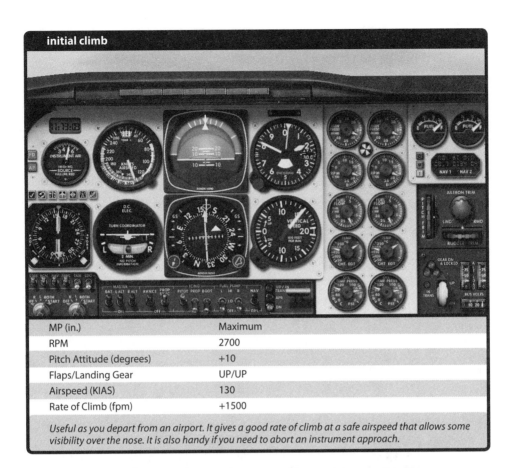

initial climb

MP (in.)	Maximum
RPM	2700
Pitch Attitude (degrees)	+10
Flaps/Landing Gear	UP/UP
Airspeed (KIAS)	130
Rate of Climb (fpm)	+1500

Useful as you depart from an airport. It gives a good rate of climb at a safe airspeed that allows some visibility over the nose. It is also handy if you need to abort an instrument approach.

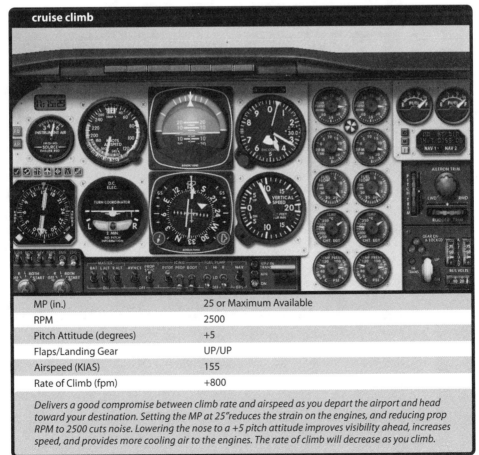

cruise climb

MP (in.)	25 or Maximum Available
RPM	2500
Pitch Attitude (degrees)	+5
Flaps/Landing Gear	UP/UP
Airspeed (KIAS)	155
Rate of Climb (fpm)	+800

Delivers a good compromise between climb rate and airspeed as you depart the airport and head toward your destination. Setting the MP at 25"reduces the strain on the engines, and reducing prop RPM to 2500 cuts noise. Lowering the nose to a +5 pitch attitude improves visibility ahead, increases speed, and provides more cooling air to the engines. The rate of climb will decrease as you climb.

flying
the aircraft
used in the
practice
flights

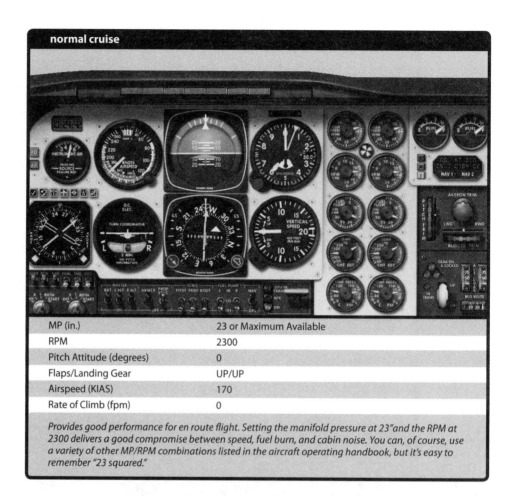

normal cruise

MP (in.)	23 or Maximum Available
RPM	2300
Pitch Attitude (degrees)	0
Flaps/Landing Gear	UP/UP
Airspeed (KIAS)	170
Rate of Climb (fpm)	0

Provides good performance for en route flight. Setting the manifold pressure at 23" and the RPM at 2300 delivers a good compromise between speed, fuel burn, and cabin noise. You can, of course, use a variety of other MP/RPM combinations listed in the aircraft operating handbook, but it's easy to remember "23 squared."

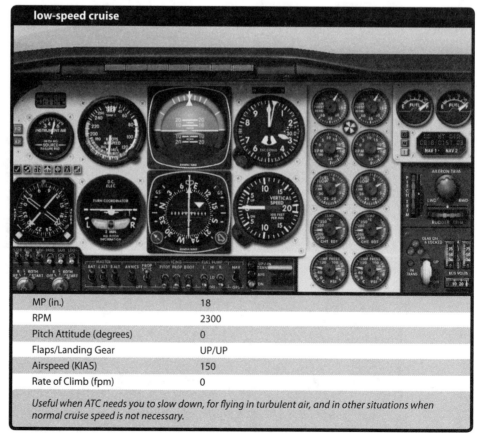

low-speed cruise

MP (in.)	18
RPM	2300
Pitch Attitude (degrees)	0
Flaps/Landing Gear	UP/UP
Airspeed (KIAS)	150
Rate of Climb (fpm)	0

Useful when ATC needs you to slow down, for flying in turbulent air, and in other situations when normal cruise speed is not necessary.

en route descent

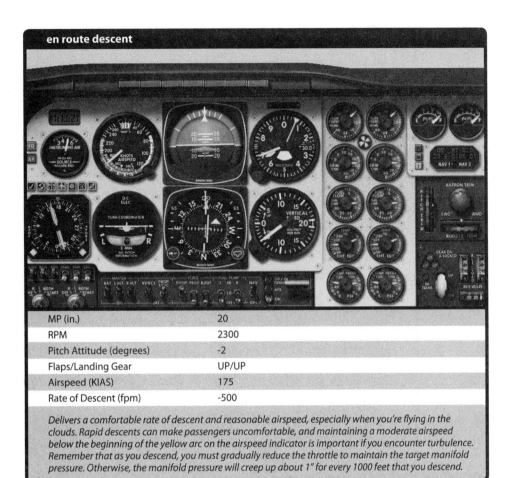

MP (in.)	20
RPM	2300
Pitch Attitude (degrees)	-2
Flaps/Landing Gear	UP/UP
Airspeed (KIAS)	175
Rate of Descent (fpm)	-500

Delivers a comfortable rate of descent and reasonable airspeed, especially when you're flying in the clouds. Rapid descents can make passengers uncomfortable, and maintaining a moderate airspeed below the beginning of the yellow arc on the airspeed indicator is important if you encounter turbulence. Remember that as you descend, you must gradually reduce the throttle to maintain the target manifold pressure. Otherwise, the manifold pressure will creep up about 1" for every 1000 feet that you descend.

expedited descent

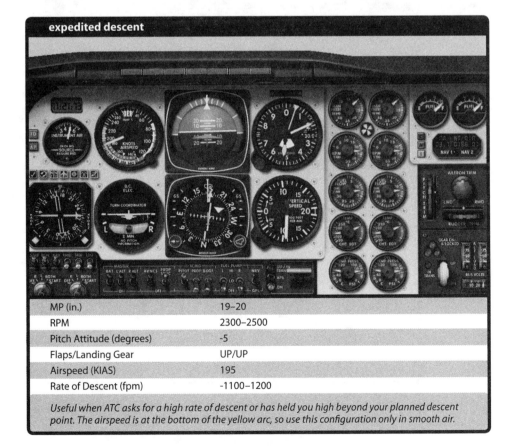

MP (in.)	19–20
RPM	2300–2500
Pitch Attitude (degrees)	-5
Flaps/Landing Gear	UP/UP
Airspeed (KIAS)	195
Rate of Descent (fpm)	-1100–1200

Useful when ATC asks for a high rate of descent or has held you high beyond your planned descent point. The airspeed is at the bottom of the yellow arc, so use this configuration only in smooth air.

flying
the aircraft
used in the
practice
flights

transition speed—level flight

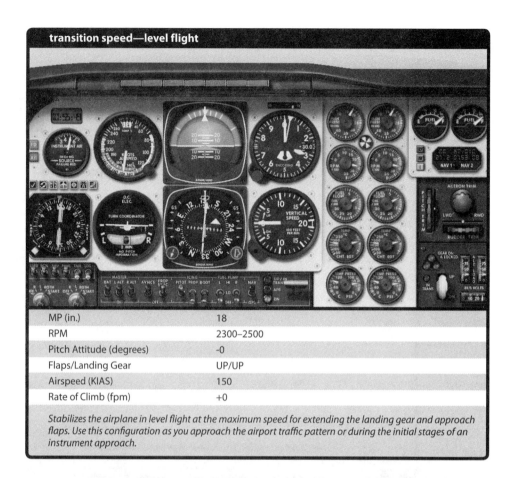

MP (in.)	18
RPM	2300–2500
Pitch Attitude (degrees)	-0
Flaps/Landing Gear	UP/UP
Airspeed (KIAS)	150
Rate of Climb (fpm)	+0

Stabilizes the airplane in level flight at the maximum speed for extending the landing gear and approach flaps. Use this configuration as you approach the airport traffic pattern or during the initial stages of an instrument approach.

level flight in the traffic pattern

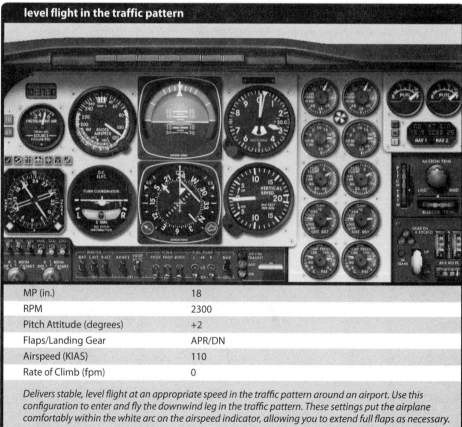

MP (in.)	18
RPM	2300
Pitch Attitude (degrees)	+2
Flaps/Landing Gear	APR/DN
Airspeed (KIAS)	110
Rate of Climb (fpm)	0

Delivers stable, level flight at an appropriate speed in the traffic pattern around an airport. Use this configuration to enter and fly the downwind leg in the traffic pattern. These settings put the airplane comfortably within the white arc on the airspeed indicator, allowing you to extend full flaps as necessary.

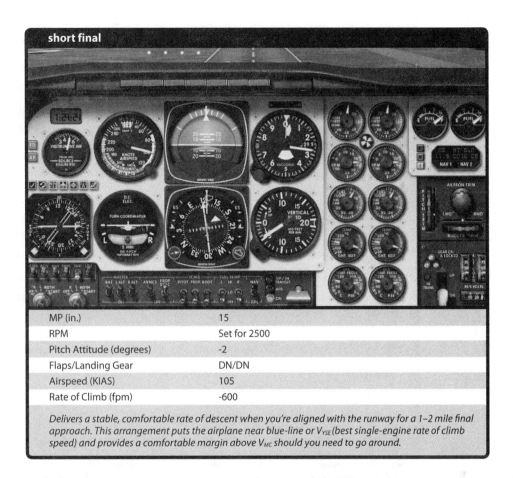

short final

MP (in.)	15
RPM	Set for 2500
Pitch Attitude (degrees)	-2
Flaps/Landing Gear	DN/DN
Airspeed (KIAS)	105
Rate of Climb (fpm)	-600

Delivers a stable, comfortable rate of descent when you're aligned with the runway for a 1–2 mile final approach. This arrangement puts the airplane near blue-line or V_{YSE} (best single-engine rate of climb speed) and provides a comfortable margin above V_{MC} should you need to go around.

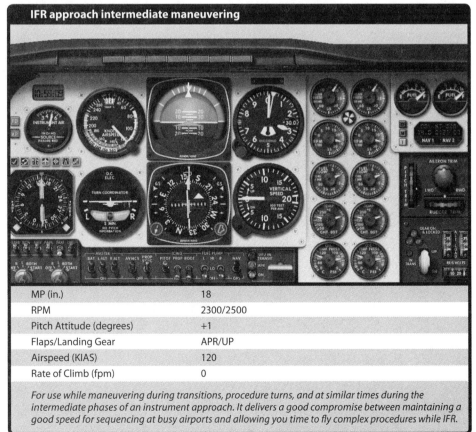

IFR approach intermediate maneuvering

MP (in.)	18
RPM	2300/2500
Pitch Attitude (degrees)	+1
Flaps/Landing Gear	APR/UP
Airspeed (KIAS)	120
Rate of Climb (fpm)	0

For use while maneuvering during transitions, procedure turns, and at similar times during the intermediate phases of an instrument approach. It delivers a good compromise between maintaining a good speed for sequencing at busy airports and allowing you time to fly complex procedures while IFR.

flying
the aircraft
used in the
practice
flights

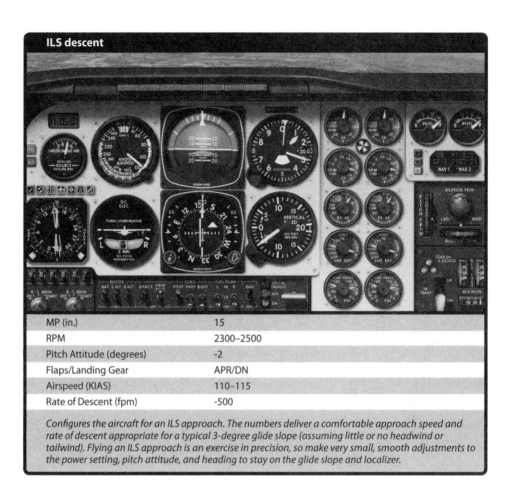

ILS descent

MP (in.)	15
RPM	2300–2500
Pitch Attitude (degrees)	-2
Flaps/Landing Gear	APR/DN
Airspeed (KIAS)	110–115
Rate of Descent (fpm)	-500

Configures the aircraft for an ILS approach. The numbers deliver a comfortable approach speed and rate of descent appropriate for a typical 3-degree glide slope (assuming little or no headwind or tailwind). Flying an ILS approach is an exercise in precision, so make very small, smooth adjustments to the power setting, pitch attitude, and heading to stay on the glide slope and localizer.

nine

supplemental information and web links

checklist

☐ official website

☐ website for this book

☐ FAA handbooks

☐ charts

☐ CD contents

☐ list of FAA handbook excerpts on the CD

To get the greatest benefit from the Practice Flights associated with this book, you should prepare thoroughly for each flight, as if getting ready for a real lesson. Chapter 1, "About this Book," included a general introduction to many resources—most available free via the web—which can help you accomplish that goal. This chapter offers more details about those resources.

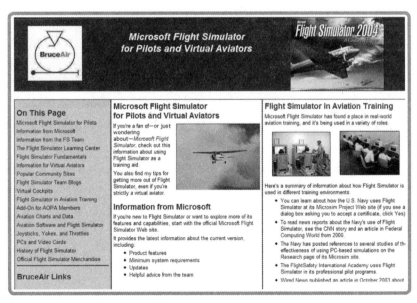

The website for this book, www.BruceAir.com, includes links to information about Flight Simulator.

official Microsoft Flight Simulator website

To learn more about *Microsoft Flight Simulator*, including its features and detailed system requirements, visit the official *Microsoft Flight Simulator* website at **http://fsinsider.com**. The official website is always the best source of the latest reliable information about Flight Simulator. You'll also find links to how-to articles and technical support that will help you get the most out of Flight Simulator. If you're a real-world pilot interested in what Flight Simulator has to offer, I recommend the article "A Real-World Pilot's Guide to Flight Simulator."

the website for this book

I have consolidated the resources associated with this book and links to other aviation-related sites that I find useful on my website. Putting all the web-based resources in one place makes it easier to keep the Practice Flights and web links up-to-date. To visit all of those resources you need to remember only one web address: **www.BruceAir.com**. You'll always find information and useful links in these general categories:

- Resources related to *Microsoft Flight Simulator*, including information about where you can find add-on aircraft and other accessories for Flight Simulator.
- Expanded information about and links associated with the topics in this book, including aviation references, training manuals, and other learning resources, many of which are free to download.
- The latest set of the Practice Flights for *Microsoft Flight Simulator* described later in this book.

FAA aviation handbooks

The briefings for the Practice Flight refer to several official FAA training handbooks. The complete manuals are available for download in Adobe Reader format (.pdf) from the FAA website. These books are comprehensive, practical guides to the core knowledge every pilot should possess. The following sections describe the specific titles mentioned in the briefings for the Practice Flights.

Pilot's Handbook of Aeronautical Knowledge (FAA-H-8083-25)

This book is primarily for aviators training for a private pilot certificate and for pilots who want to refresh their knowledge of flying fundamentals. It introduces pilots to the theory and practical knowledge they will need as they progress in their pilot training from aerodynamics, engines, flight instruments and navigation to avionics, weather, and airspace.

Pilot's Handbook of Aeronautical Knowledge (FAA 8083-25)

Contents

Chapter 1—Aircraft Structure
Chapter 2—Principles of Flight
Chapter 3—Aerodynamics of Flight
Chapter 4—Flight Controls
Chapter 5—Aircraft Systems
Chapter 6—Flight Instruments

Chapter 7—Flight Manuals and Other Documents
Chapter 8—Weight and Balance
Chapter 9—Aircraft Performance
Chapter 10—Weather Theory
Chapter 11—Weather Reports, Forecasts, and Charts
Chapter 12—Airport Operations
Chapter 13—Airspace
Chapter 14—Navigation
Chapter 15—Aeromedical Factors
Chapter 16—Aeronautical Decision Making

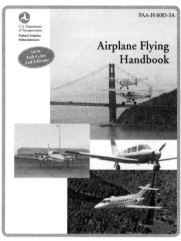

Airplane Flying Handbook (FAA-H-8083-3A)

This book is the hands-on, how-to guide to flying all the maneuvers required for the private pilot and commercial pilot certificates. It is intended for student pilots who are learning to fly and for current pilots and flight instructors who want to expand their knowledge and skills. It is one of the official references that pilot examiners use when giving checkrides to pilot applicants.

Instrument Flying Handbook
(FAA-H-8083-3A)

Contents

Chapter 1—Introduction to Flight Training
Chapter 2—Ground Operations
Chapter 3—Basic Flight Maneuvers
Chapter 4—Slow Flight, Stalls, and Spins
Chapter 5—Takeoff and Departure Climbs
Chapter 6—Ground Reference Maneuvers
Chapter 7—Airport Traffic Patterns
Chapter 8—Approaches and Landings
Chapter 9—Performance Maneuvers
Chapter 10—Night Operations
Chapter 11—Transition to Complex Airplanes
Chapter 12—Transition to Multiengine Airplanes
Chapter 13—Transition to Tailwheel Airplanes
Chapter 14—Transition to Turbopropeller Powered Airplanes
Chapter 15—Transition to Jet Powered Airplanes
Chapter 16—Emergency Procedures
Glossary

supplemental information and web links

Instrument Flying Handbook (FAA-H-8083-15)

This is the FAA's guide to the aeronautical knowledge required to operate in instrument meteorological conditions (IMC). It is intended for pilots training for an instrument rating and for the instrument flight instructors who teach them. The *Instrument Flying Handbook* includes both theory and practical guidance on how to operate an aircraft solely by reference to instruments, from basic attitude instrument flying and working with ATC to flying DME arcs, holding patterns, and instrument approaches.

Contents

Chapter 1—Human Factors
Chapter 2—Aerodynamic Factors
Chapter 3—Flight Instruments
Chapter 4—Airplane Attitude Instrument Flying
Chapter 5—Airplane Basic Flight Maneuvers
Chapter 6—Helicopter Attitude Instrument Flying
Chapter 7—Navigation Systems
Chapter 8—The National Airspace System
Chapter 9—The Air Traffic Control System
Chapter 10—IFR Flight
Chapter 11—Emergency Operations
Appendices
Glossary
Index

Instrument Flying Handbook
(FAA-H-8083-15)

Instrument Procedures Handbook (FAA-H-8261-1)

This handbook is a practical guide and supplement to the *Instrument Flying Handbook* for all instrument pilots. It covers all the phases of instrument flight and provides detailed information about advanced IFR procedures and equipment.

Contents

Chapter 1—IFR Operations in the National
   Airspace System
Chapter 2—Takeoffs and Departures
Chapter 3—En Route Operations
Chapter 4—Arrivals
Chapter 5—Approaches
Chapter 6—System Improvement Plans
Appendix A—Airborne Navigation Databases
Appendix B—Approach Chart Format Changes
Appendix C—Helicopter Instrument Procedures
Appendix D—Acronyms and Glossary
Index

Instrument Procedures Handbook
(FAA-H-8261-1)

additional FAA references

In addition to the training handbooks described above, the FAA offers many more re-sources via the web, including Advisory Circulars, *FAA Aviation News*, and guides to op-erational information. The following publications are especially helpful as you use the Practice Flights:

- *Aeronautical Chart User's Guide* from the FAA National Aeronautical Charting Office (NACO). Explains the symbols and terminology used on all U.S. government-issue VFR and IFR charts.
- *Aeronautical Information Manual (AIM)*. The AIM is the FAA's "Official Guide to Basic Flight Information and ATC Procedures." The AIM is available online in both Adobe Reader (.pdf) and web (HTML) formats.
- *Private Pilot: Practical Test Standards for Airplane* (FAA-S-8081-14)
- *Instrument Rating: Practical Test Standards* (FAA-S-8081-4)

ASA publishes reprints of many official FAA handbooks and references.

charts and other references

Charts, diagrams, and other references required to use specific Practice Flights are in-cluded in the briefings available for download from the website for this book, where you'll also find links to sources for additional charts, including web-based aviation maps and flight-planning tools.

FAA NACO charts

The FAA National Aeronautical Charting Office publishes VFR charts (Sectionals and WACs), the Airport/Facility Directory (A/FD), airport diagrams, and IFR charts (en route, departure, arrival, and approach).

At present, only the charts for IFR procedures are available as free downloads from NACO. However, several websites provide online access to Sectional charts, and you can order complete sets of VFR charts in electronic form from several providers, including NACO.

supplemental information and web links

online VFR charts and IAP charts

The website for this book includes links to several websites that feature online versions of Sectional, Class B, and instrument procedure charts. You can use these sites to prepare for the VFR and IFR Practice Flights.

Jeppesen charts

Jeppesen does not make its charts available for free download from the web. However, you can order sets of charts from Jeppesen or pilot shops. For the latest information about the products Jeppesen offers, visit the Jeppesen website.

If you use Jeppesen charts, you can learn all about them by downloading the free Chart Clinic Reprints available on the "Aviation Resources" section of the Jeppesen website.

Jeppesen produced the Chart Clinics series that appeared in AOPA Pilot magazine from 1998 through 2001. Each two-page installment in the series includes a wealth of information about Jeppesen departure, en route, arrival and approach charts, plus detailed explanations of IFR procedures and terminology. These detailed guides are available for free download in PDF format from the Jeppesen website.

Unfortunately, the Chart Clinic articles are organized by date of original publication, not by topic; but the following table points you to specific articles.

chart clinic topic	date of publication
Reporting points, navigation aids, intersections, etc.	April 1998
Airspace	July 1998
Airways, en route altitudes, etc.	September 1998
Plotting airway and direct routes	November 1998
More about minimum altitudes	December 1998
Airway symbology	January 1999
Understanding the title/index section of approach charts	February 1999
The Briefing Strip	March 1999
More about communications	March 1999
The plan view	June 1999
The segments of an approach	July 1999, August 1999, September 1999
The profile view	October 1999
Non-precision approaches	November 1999, December 1999
Minimums	January 2000, February 2000
More on minimums and missed approaches	March 2000
Airport diagrams	May 2000
Departure procedures	June 2000
Arrival procedures	July 2000

Note that some symbols and terminology used on Jeppesen charts—especially those related to GPS navigation and approaches—have changed since the Chart Clinic series was published. For the latest information, refer to current Jeppesen legends and glossaries and the *Instrument Procedures Guide* (JS312407), the Jeppesen edition of the FAA *Instrument Procedures Handbook*, available from Jeppesen and bookstores.

CD contents

The CD that accompanies this book includes the following resources:

- Documents in Adobe Reader format (.pdf) with the latest information about how to install the Practice Flights, corrections and additions to the text of this book, and a list of useful web links.
- The VFR and IFR Practice Flights (each Practice Flight includes .flt and .wx files that contain the initial conditions and weather respectively).
- A complete list of these Practice Flights.
- Preflight Briefings for the Practice Flights, including references to charts where appropriate.
- Excerpts from the FAA training handbooks mentioned in this chapter and in the Preflight Briefings for the Practice Flights (see list starting on the next page).
- Quick-reference guides to key features in Flight Simulator that are helpful while using the Practice Flights.
- A checklist to help you create your own Flights.

The following diagram shows the basic folder structure of the CD.

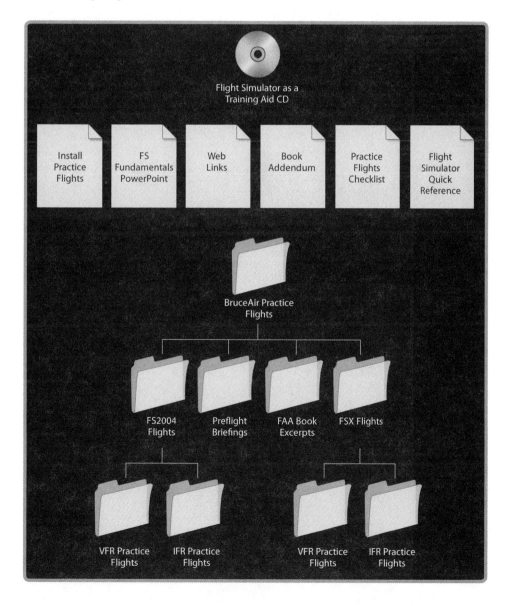

excerpts from FAA training handbooks and sources

The following list shows most of the excerpts from FAA training handbooks and sources that are included on the CD for this book. The "Background Information" section for each category of Practice Flight includes references to these excerpts. All of the documents are in Adobe Reader format (.pdf). As noted earlier, you can download complete copies of these resources from the FAA website.

Pilot's Handbook of Aeronautical Knowledge (FAA-H-8083-25)

PHAK-Chap01_AircraftStructure.pdf
PHAK-Chap04_FlightControls.pdf
PHAK-Chap05_AircraftSystems.pdf
PHAK-Chap06_FlightInstruments.pdf
PHAK-Chap12_AirportOperations.pdf
PHAK-Chap13_Airspace.pdf
PHAK-Chap14_Navigation.pdf
PHAK-Chap14_NDB.pdf
PHAK-Chap14_VOR.pdf
PHAK-Glossary.pdf

Airplane Flying Handbook (FAA-H-8083-3A)

AFH-Chap04_SlowFlightStallsSpins.pdf
AFH-Chap06_TakeoffsDeparture.pdf
AFH-Chap07_AirportTrafficPatterns.pdf
AFH-Chap08_ApproachLanding.pdf
AFH-Chap10_NightOps.pdf
AFH-Chap11_ComplexAirplanes.pdf
AFH-Chap16_Emergencies.pdf
AFH-Glossary.pdf

Instrument Flying Handbook (FAA-H-8083-15)

IFH-Chap03_FlightInstruments.pdf
IFH-Chap04_AirplaneAttitudeInstrumentFlying.pdf
IFH-Chap05_BasicIFRPatterns.pdf
IFH-Chap07_DME.pdf
IFH-Chap07_GPS.pdf
IFH-Chap07_ILS.pdf
IFH-Chap07_NDB.pdf
IFH-Chap07_VOR.pdf
IFH-Chap08_EnRouteProcedures.pdf
IFH-Chap08_ProcedureTurn.pdf
IFH-Chap10_ApproachVariations.pdf
IFH-Chap10_DPs.pdf
IFH-Chap10_HoldingProcedures.pdf
IFH-App01_ClearanceShorthand.pdf
IFH-App02_IFRSyllabus.pdf
IFH-Glossary.pdf

Instrument Procedures Handbook (FAA-H-8261-1)

IPH-Chap02_DPs.pdf
IPH-Chap03_En_Route.pdf
IPH-Chap03_HoldingProcedures.pdf
IPH-Chap05_DME-Arcs.pdf
IPH-Chap05_GPS.pdf
IPH-Chap05_ILS.pdf
IPH-Chap05_NDB_Approach.pdf
IPH-Chap05_ProcedureTurns.pdf
IPH-Chap05_VOR-Approach.pdf
IPH-Glossary.pdf

Aeronautical Information Manual

AIM.pdf
AIM-Pilot-ControllerGlossary.pdf

Practical Test Standards

FAA-S-8081-12B_Comm-PTS.pdf
FAA-S-8081-14A_Private-PTS.pdf
IFR-PTS-FAA-S-8081-4D.pdf

Aeronautical Chart User's Guide

NACO-ACUG_7th_VFR_Intro.pdf
NACO-ACUG_7th_VFR_Symbols.pdf
NACO-ACUG_7th_IFR_EnRouteIntro.pdf
NACO-ACUG_7th_IFR_EnRouteSymbols.pdf
NACO-ACUG_7th_IAP_Intro.pdf
NACO-ACUG_7th_IAP_Symbols.pdf

ten

introduction
to the VFR
practice flights

checklist

☐ effective use

☐ practice flight
   aircraft, locations,
   and categories

☐ basic flying skills

☐ using radios

☐ traffic patterns

☐ takeoffs and
   landings

☐ maneuvers

☐ VOR, NDB, GPS

☐ basic instrument
   flying

☐ night flying

☐ marginal weather

☐ emergencies

☐ commercial pilot
   practice flights

This chapter describes the categories of VFR Practice Flights available on the CD and from the website for this book. The complete, current list of VFR Practice Flights is available on the website for this book. For background information about the Flights featured in *Microsoft Flight Simulator* and the Practice Flights, see Chapter 7, "About the Practice Flights."

These Practice Flights support lessons associated with the Private Pilot and Commercial Pilot Practical Test Standards (PTS) published by the FAA. Note that the Performance Standards listed for each Practice Flight do not include all of the standards and evaluation criteria from the respective official test standards. For the complete set of applicable criteria, refer to the latest official versions of the PTS. I have also included Practice Flights that focus on areas not specifically addressed by flying tasks in the test standards—for example, operations in marginal weather. Of course, you can use the Practice Flights in any order you choose.

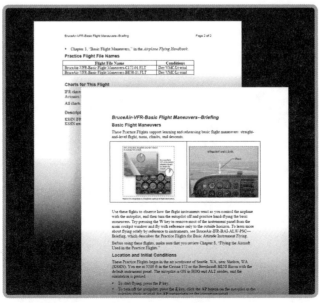

Briefing for a typical VFR Practice Flight

most effective use

For the reasons discussed in Chapter 2, "Using Flight Simulator as a Training Aid," and Chapter 3, "Best Practices for Using Flight Simulator," many VFR Practice Flights are best used as demonstration tools, especially before a student has developed basic flying skills in the airplane. For example, the Practice Flights that support basic flight maneuvers such as slow flight and stalls can help an instructor teach, and a student learn, the proper procedures for configuring the aircraft and performing the maneuvers according to the PTS. The Practice Flights can even help a student learn the pitch attitudes, airspeeds, and power settings associated with such maneuvers. Still, operating a PC-based simulator at a desk cannot refine the feel for real control forces, coordination of rudder, and other elements of flying a real aircraft.

aircraft used in the VFR practice flights

The VFR Practice Flights use the Cessna 172SP Skyhawk and Beechcraft BE58 Baron. For more information about the aircraft, see Chapter 8, "Flying the Aircraft Used in the Practice Flights." You can find additional information about these airplanes, including checklists and flying hints, in the "Aircraft Information" section of the Learning Center.

locations

Most of the VFR Practice Flights begin in the Pacific Northwest. The preflight briefing for each Practice Flight indicates the specific startling location for that Practice Flight.

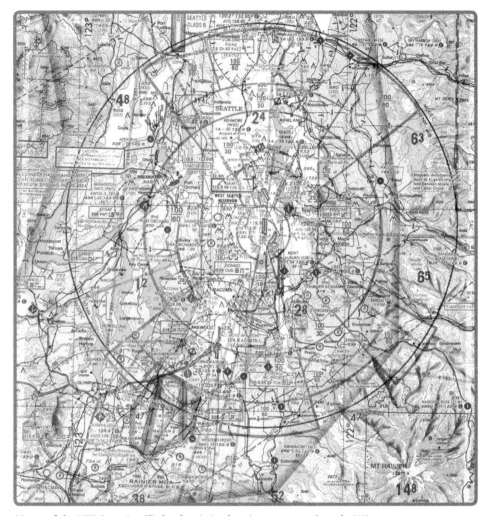

Many of the VFR Practice Flights begin in the airspace near Seattle, WA.

categories

The explanation of each category of VFR Practice Flights includes:

The purpose of the Practice Flights included in the category.

- References to the background information from FAA training handbooks, the Aeronautical Information Manual, and other resources that you should review before flying Practice Flights in that category.
- Excerpts from the Performance Standards of the relevant FAA PTS to help you evaluate flights.
- Suggestions and tips for how to use the Practice Flights most effectively in a given category. These appy to students, pilots and flight instructors.

The preflight briefing associated with each Practice Flight provides additional information, including:

- The starting location of the flight.
- The weather, time of day, and other pertinent information about the environment for the flight.
- Any specific instructions for the flight.
- A chart or chart excerpt that covers the area in which that Practice Flight takes place.

You can find examples of typical briefings for the VFR Practice Flights in Chapter 11, "Briefings for VFR Practice Flights."

private pilot practice flights

These are covered first in this chapter and they use the Cessna 172SP Skyhawk, focusing on the following areas of operation from the Private Pilot PTS (plus supplemental operations that may be useful in a typical training program):

Basic Flying Skills
Using Radios
Airport Operations—Traffic Patterns
Takeoffs, Landings, and Go-Arounds
Performance Maneuver—Steep Turns
Navigation
Slow Flight and Stalls
Basic Instrument Maneuvers
Night Operations
Flight in Marginal Weather
Emergencies

commercial pilot practice flights

As explained later in this chapter, the Commercial Pilot Practice Flights use the Beechcraft BE58 Baron and focus on the operation of complex/high-performance aircraft.

sources of background information

Excerpts from public-domain resources such as the FAA training handbooks are provided in Adobe Reader format (.pdf) on the CD and website for this book. For detailed descriptions of the resources, see Chapter 1, "About this Book" and Chapter 9, "Supplemental Information."

basic flying skills

For Private Pilot practice, use the Practice Flights in this section to observe and practice basic flying skills, including:

- The purpose and effect of the primary flight and engine controls.
- Purpose and function of the basic flight instruments.
- Operation of the engine controls and purpose and function of the engine instruments.
- Use of elevator trim.

- Configuring the aircraft for various situations by applying the Power + Configuration/Attitude = Performance equation as described in Chapter 8, "Flying the Aircraft Used in the Practice Flights" and in Chapter 3, "Basic Flight Maneuvers" in the *Airplane Flying Handbook*.
- Correlating information from instruments and visual flying according to the principle of Integrated Flight Instruction, as described in "Basic Flight Maneuvers" in the *Airplane Flying Handbook*.
- You can use the Baron version of this Practice Flight to introduce operating an aircraft with a constant-speed propeller, retractable landing gear, and advanced instruments like the HSI.

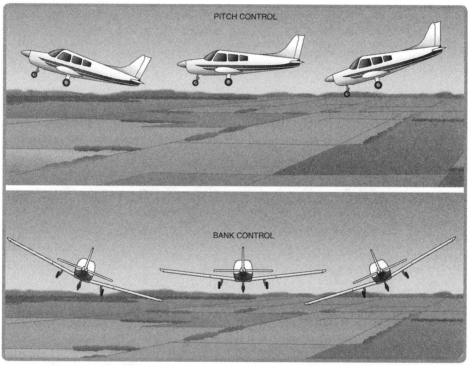

Airplane attitude is based on relative position of the nose and wings on the natural horizon.
(from FAA-H-8083-3A)

background information

Before using "Basic Flying Skills" Practice Flights, you should be familiar with the following background information:

Chapter 4: Flight Controls, in the *Pilot's Handbook of Aeronautical Knowledge*.

Chapter 6: Flight Instruments, *Pilot's Handbook of Aeronautical Knowledge*.

Chapter 3: Basic Flight Maneuvers, in the *Airplane Flying Handbook*.

performance standards

The general standards established in the FAA Private Pilot PTS for operating the aircraft through normal visual maneuvers are:

- Maintain altitude ±100 feet
- Hold or roll out on heading ±10°
- Maintain specified bank ±5°
- Maintain appropriate airspeed ±10 knots

suggestions and tips for basic flying skills practice flights

- Use the concept of "the numbers" described in Chapter 8, "Flying the Aircraft Used in the Practice Flights," to help you establish and maintain stable configurations.
- Associate specific pitch and bank attitudes, power settings, and aircraft configurations with aircraft performance.
- Use the autopilot at first so that you can observe how the instruments react as the airplane turns, climbs, and descends.

using radios

Use the Practice Flights in this section to learn and practice procedures associated with radio communications, including:

- Locating frequencies and other information about ATC facilities from charts and other sources.
- Tuning and configuring radios in a typical avionics stack.

background information

Before using these Practice Flights, you should be familiar with the following background information:

The Safety Advisors "Operations at Towered Airports," "Operations at Nontowered Airports," and "Say Intentions…When you need ATC's help," available in the free online library at the AOPA Air Safety Foundation website.

AIM Section 4–2 "Radio Communications Phraseology and Techniques."

Pilot/Controller Glossary addendum in AIM.

performance standards

These Practice Flights apply to Task III.A. "Radio Communications and ATC Light Signals" in the FAA Private Pilot PTS, which include the following criteria:

Objective—Determine that the pilot:

- Exhibits knowledge of the elements related to radio communications and ATC light signals.
- Selects appropriate frequencies.
- Transmits using recommended phraseology.
- Acknowledges radio communications and complies with instructions.

suggestions and tips for "using radios" practice flights

- Follow the process described in the section "Flight Simulator: A 'Swiss Army Knife'" in Chapter 2, "Using Flight Simulator as a Training Aid," to teach and practice the basics of using the communications and navigation radios in a typical avionics stack.
- Use the mouse and mouse wheel as described in Chapter 3, "Flight Simulator Essentials," to operate the controls on the radios.

airport operations—traffic patterns

Use the Practice Flights in this section to observe and practice operations in the airport traffic pattern, including:

- Gathering information about airports from a sectional chart and applying that information to entering and flying traffic patterns at towered and non-towered airports.
- Orientation of the aircraft with respect to standard, non-standard, and abbreviated traffic patterns.
- Configuring the aircraft for various situations by applying the Power + Configuration/Attitude = Performance equation as described in Chapter 8, "Flying the Aircraft Used in the Practice Flights."
- Integrating cockpit flow checks/checklists with operations in the vicinity of an airport.
- Setting up the communications radios to receive ATIS and ASOS reports and to communicate with a control tower or on a CTAF, as appropriate.

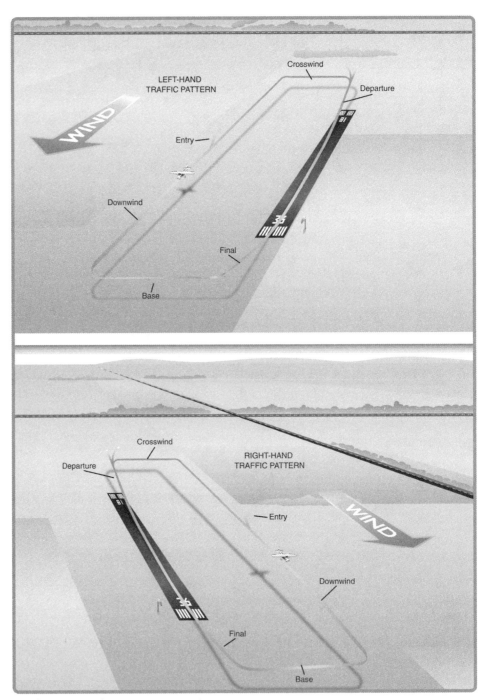

Traffic patterns *(from FAA-H-8083-3A)*

background information

Before using "Airport Operations—Traffic Patterns" Practice Flights, you should be familiar with the following background information:

Chapter 12: Airport Operations, in the *Pilot's Handbook of Aeronautical Knowledge.*

Chapter 7: Airport Traffic Patterns, in the *Airplane Flying Handbook.*

The Safety Advisors "Operations at Towered Airports" and "Operations at Nontowered Airports" available in the free online library at the AOPA Air Safety Foundation website.

Advisory Circulars 90-42F, "Traffic Advisory Practices at Airports Without Operating Control Towers" and 90-66A "Recommended Standard Traffic Patterns and Practices for Aeronautical Operations at Airports without Operating Control Towers."

performance standards

These Practice Flights apply to Task III. B., "Airport Operations: Traffic Patterns" in the FAA Private Pilot PTS, which include:

Objective—Determine that the pilot:

- Exhibits knowledge of the elements related to traffic patterns. This shall include procedures at airports with and without operating control towers, prevention of runway incursions, collision avoidance, wake turbulence avoidance, and wind shear.
- Complies with proper traffic pattern procedures.
- Corrects for wind drift to maintain the proper ground track.
- Maintains orientation with the runway/landing area in use.
- Maintains traffic pattern altitude, ±100 feet, and the appropriate airspeed, ±10 knots.

suggestions and tips for airport operations—traffic patterns practice flights

- Use the concept of "the numbers" described in Chapter 8, "Flying the Aircraft Used in the Practice Flights," to help you establish and maintain stable configurations.
- Fly the traffic patterns at a variety of speeds and configurations to practice fitting into to the flow in a busy traffic pattern.
- Use the views described in Chapter 4, "Flight Simulator Essentials," to look around as the airplane progresses through the legs of a standard traffic pattern.

takeoffs, landings, and go-arounds

Use the Practice Flights in this section to observe and practice skills and procedures associated with takeoffs and landings, including:

- The procedures for normal, soft-field, and short-field takeoffs and landings.
- Configuring the aircraft for various situations by applying the Power + Configuration/Attitude = Performance equation.
- Developing the appropriate visual cues for judging approach angle and touchdown point.
- Observing the effects of wind drift during departures and approaches.

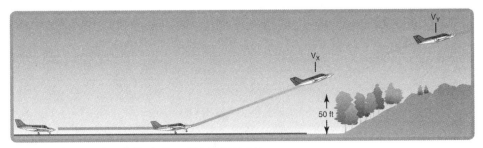

Short-field takeoff and climb (from FAA-H-8083-3A)

background information

Before using the "Takeoffs, Landings, and Go-Arounds" Practice Flights, you should be familiar with the following background information:

Chapter 5: Takeoff and Departure Climbs, in the *Airplane Flying Handbook*.

Chapter 8: Approaches and Landings, in the *Airplane Flying Handbook*.

The Safety Advisor "Ups and Downs of Takeoffs and Landings" available in the free online library at the AOPA Air Safety Foundation website.

performance standards

These Practice Flights apply to Task IV, "Takeoffs, Landings, and Go-Arounds" in the FAA Private Pilot PTS, which include the following criteria:

normal and crosswind takeoff and climb

Objective—Determine that the pilot:

- Exhibits knowledge of the elements related to a normal and crosswind takeoff, climb operations, and rejected takeoff procedures.
- Positions the flight controls for the existing wind conditions.
- Clears the area; taxies into the takeoff position and aligns the airplane on the runway center/takeoff path.
- Advances the throttle smoothly to takeoff power.
- Establishes and maintains the most efficient attitude.
- Lifts off at the recommended airspeed and accelerates to V_Y.
- Establishes a pitch attitude that will maintain V_Y +10/−5 knots.
- Retracts the landing gear, if appropriate, and flaps after a positive rate of climb is established.
- Maintains takeoff power and V_Y +10/−5 knots to a safe maneuvering altitude.
- Maintains directional control and proper wind-drift correction throughout the takeoff and climb.
- Complies with noise abatement procedures.
- Completes the appropriate checklist.

normal and crosswind approach and landing

Objective—Determine that the pilot:

- Exhibits knowledge of the elements related to a normal and crosswind approach and landing.
- Considers the wind conditions, landing surface, obstructions, and selects a suitable touchdown point.
- Establishes the recommended approach and landing configuration and airspeed, and adjusts pitch attitude and power as required.
- Maintains a stabilized approach and recommended airspeed, or in its absence, not more than 1.3 V_{SO}, +10/-5 knots, with wind gust factor applied.
- Makes smooth, timely, and correct control application during the roundout and touchdown.
- Touches down smoothly at approximate stalling speed.
- Touches down at or within 400 feet (120 meters) beyond a specified point, with no drift, and with the airplane's longitudinal axis aligned with and over the runway center/landing path.

- Maintains crosswind correction and directional control throughout the approach and landing sequence.
- Completes the appropriate checklist.

soft-field takeoff and climb

Objective—Determine that the pilot:

- Exhibits knowledge of the elements related to a soft-field takeoff and climb.
- Positions the flight controls for existing wind conditions and to maximize lift as quickly as possible.
- Clears the area; taxies onto the takeoff surface at a speed consistent with safety without stopping while advancing the throttle smoothly to takeoff power.
- Establishes and maintains a pitch attitude that will transfer the weight of the airplane from the wheels to the wings as rapidly as possible.
- Lifts off at the lowest possible airspeed and remains in ground effect while accelerating to V_X or V_Y as appropriate.
- Establishes a pitch attitude for V_X or V_Y, as appropriate, and maintains selected airspeed +10/−5 knots, during the climb.
- Retracts the landing gear, if appropriate, and flaps after clear of any obstacles or as recommended by the manufacturer.
- Maintains takeoff power and V_X or V_Y +10/−5 knots to a safe maneuvering altitude.
- Maintains directional control and proper wind-drift correction throughout the takeoff and climb.
- Completes the appropriate checklist.

soft-field approach and landing

Objective—Determine that the pilot:

- Exhibits knowledge of the elements related to a soft-field approach and landing.
- Considers the wind conditions, landing surface and obstructions, and selects the most suitable touchdown area.
- Establishes the recommended approach and landing configuration, and airspeed; adjusts pitch attitude and power as required.
- Maintains a stabilized approach and recommended airspeed, or in its absence not more than 1.3 V_{SO} +10/−5 knots, with wind gust factor applied.
- Makes smooth, timely, and correct control application during the roundout and touchdown.
- Touches down softly with no drift, and with the airplane's longitudinal axis aligned with the runway/landing path.
- Maintains crosswind correction and directional control throughout the approach and landing sequence.
- Maintains proper position of the flight controls and sufficient speed to taxi on the soft surface.
- Completes the appropriate checklist.

short-field takeoff and maximum performance climb

Objective—Determine that the pilot:

- Exhibits knowledge of the elements related to a short-field takeoff and maximum performance climb.

- Positions the flight controls for the existing wind conditions; sets the flaps as recommended.
- Clears the area; taxies into takeoff position utilizing maximum available takeoff area and aligns the airplane on the runway centerline.
- Applies brakes (if appropriate), while advancing the throttle smoothly to takeoff power.
- Establishes and maintains the most efficient lift-off attitude.
- Lifts off at the recommended airspeed, and accelerates to the recommended obstacle clearance airspeed or V_x.
- Establishes a pitch attitude that will maintain the recommended obstacle clearance airspeed, or V_x, +10/−5 knots, until the obstacle is cleared, or until the airplane is 50 feet above the surface.
- After clearing the obstacle, establishes the pitch attitude for V_y, accelerates to V_y, and maintains V_y, +10/−5 knots, during the climb.
- Retracts the landing gear, if appropriate, and flaps after clear of any obstacles or as recommended by manufacturer.
- Maintains takeoff power and V_y +10/−5 to a safe maneuvering altitude.
- Maintains directional control and proper wind-drift correction throughout the takeoff and climb.
- Completes the appropriate checklist.

short-field approach and landing

Objective—Determine that the pilot:

- Exhibits knowledge of the elements related to a short-field approach and landing.
- Considers the wind conditions, landing surface, obstructions, and selects the most suitable touchdown point.
- Establishes the recommended approach and landing configuration and airspeed; adjusts pitch attitude and power as required.
- Maintains a stabilized approach and recommended approach airspeed, or in its absence not more than 1.3 V_{so}, +10/−5 knots, with wind gust factor applied.
- Makes smooth, timely, and correct control application during the roundout and touchdown.
- Touches down at or within 200 feet beyond a specified point, with no side drift, minimum float and with the airplane's longitudinal axis aligned with and over the runway center/landing path.
- Maintains crosswind correction and directional control throughout the approach and landing sequence.
- Applies brakes as necessary, to stop in the shortest distance consistent with safety.
- Completes the appropriate checklist.

go-around/rejected landing

Objective—Determine that the pilot:

- Exhibits knowledge of the elements related to a go-around/rejected landing.
- Makes a timely decision to discontinue the approach to landing.
- Applies takeoff power immediately and transitions to climb pitch attitude for V_y, and maintains V_y +10/−5 knots.
- Retracts the flaps as appropriate.
- Retracts the landing gear, if appropriate, after a positive rate of climb is established.
- Maneuvers to the side of the runway/landing area to clear and avoid conflicting traffic.

- Maintains takeoff power V$_Y$ +10/−5 to a safe maneuvering altitude.
- Maintains directional control and proper wind-drift correction throughout the climb.
- Completes the appropriate checklist.

suggestions and tips for takeoffs, landings, and go-arounds practice flights

- Use these Practice Flights to demonstrate and help reinforce the appropriate procedures for takeoffs and landings.
- Use the concept of "the numbers" described in Chapter 8, "Flying the Aircraft Used in the Practice Flights," to establish and maintain stable configurations.
- Use the views described in Chapter 4, "Flight Simulator Essentials," to provide a better view of the runway and to look around as the airplane progresses through the maneuver. Use the outside "spot plane" view to correlate the picture from the pilot's seat with the aircraft's attitude.

performance maneuver—steep turns

Use the Practice Flights in this section to observe and practice steep turns, including:

- The procedures for setting up and flying performance maneuvers.
- Configuring the aircraft for various situations by applying the Power + Configuration/Attitude = Performance equation.

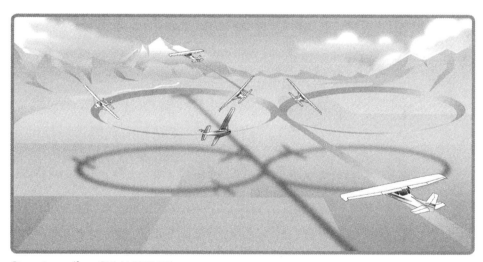

Steep turns (from FAA-H-8083-3A)

background information

Before using the "Performance Maneuver—Steep Turns" Practice Flights, you should be familiar with the following background information:

Chapter 9: Performance Maneuvers, in the *Airplane Flying Handbook*.

performance standards

These Practice Flights apply to Task V, "Performance Maneuver—Steep Turns" in the FAA Private Pilot PTS, which include the following criteria:

Objective—Determine that the pilot:

- Exhibits knowledge of the elements related to steep turns.
- Establishes the manufacturer's recommended airspeed or if one is not stated, a safe airspeed not to exceed V_A.
- Rolls into a coordinated 360° turn; maintains a 45° bank.
- Divides attention between airplane control and orientation.
- Maintains the entry altitude, ±100 feet (30 meters), airspeed, ±10 knots, bank, ±5°; and rolls out on the entry heading, ±10°.

suggestions and tips for performance maneuver—steep turns practice flights

- Use the concept of "the numbers" described in Chapter 8, "Flying the Aircraft Used in the Practice Flights," to establish and maintain stable configurations.
- Use the views described in Chapter 4, "Flight Simulator Essentials," to provide a better view of horizon. The "virtual cockpit" and "wide" views make it easier to see the visual horizon and to avoid fixating on the instrument panel.

slow flight and stalls

Use the Practice Flights in this section to learn and practice procedures associated with slow flight and stalls, including:

- Configuring the airplane for slow flight and stalls.
- The procedures for setting up and flying specific maneuvers.

Slow flight: low airspeed, high angle of attack, high power, and constant altitude (from FAA-H-8083-3A)

background information

Before using the Slow Flight and Stalls Practice Flights, you should be familiar with the following background information:

Chapter 2: Principles of Flight, *Pilot's Handbook of Aeronautical Knowledge*.

Chapter 3: Aerodynamics of Flight, *Pilot's Handbook of Aeronautical Knowledge*.

Chapter 5: Slow Flight, Stalls, and Spins, *Airplane Flying Handbook*.

performance standards

These Practice Flights apply to Task VIII "Slow Flight and Stalls" in the FAA Private Pilot PTS, which include the following criteria:

maneuvering during slow flight

Objective—Determine that the pilot:

- Exhibits knowledge of the elements related to maneuvering during slow flight.
- Selects an entry altitude that will allow the task to be completed no lower than 1,500 feet AGL.
- Establishes and maintains an airspeed at which any further increase in angle of attack, increase in load factor, or reduction in power, would result in an immediate stall.
- Accomplishes coordinated straight-and-level flight, turns, climbs, and descents with landing gear and flap configurations specified by the examiner.
- Divides attention between airplane control and orientation.
- Maintains the specified altitude, ±100 feet (30 meters); specified heading, ±10°; airspeed, +10/–0 knots; and specified angle of bank, ±10°.

power-off stalls

Objective—Determine that the pilot:

- Exhibits knowledge of the elements related to power-off stalls.
- Selects an entry altitude that allows the task to be completed no lower than 1,500 feet AGL.
- Establishes a stabilized descent in the approach or landing configuration, as specified by the examiner.
- Transitions smoothly from the approach or landing attitude to a pitch attitude that will induce a stall.
- Maintains a specified heading, ±10°, in straight flight; maintains a specified angle of bank not to exceed 20°, ±10°; in turning flight, while inducing the stall.
- Recognizes and recovers promptly after the stall occurs by simultaneously reducing the angle of attack, increasing power to maximum allowable, and leveling the wings to return to a straight-and-level flight attitude with a minimum loss of altitude appropriate for the airplane.
- Retracts the flaps to the recommended setting; retracts the landing gear, if retractable, after a positive rate of climb is established.
- Accelerates to V_x or V_y speed before the final flap retraction; returns to the altitude, heading, and airspeed specified by the examiner.

power-on stalls

Objective—Determine that the pilot:

- Exhibits knowledge of the elements related to power-on stalls.
- Selects an entry altitude that allows the task to be completed no lower than 1,500 feet AGL.
- Establishes the takeoff or departure configuration. Sets power to no less than 65 percent available power.
- Transitions smoothly from the takeoff or departure attitude to the pitch attitude that will induce a stall.
- Maintains a specified heading, ±10°, in straight flight; maintains a specified angle of bank not to exceed 20°, ±10°, in turning flight, while inducing the stall.
- Recognizes and recovers promptly after the stall occurs by simultaneously reducing the angle of attack, increasing power as appropriate, and leveling the wings to return to a straight-and-level flight attitude with a minimum loss of altitude appropriate for the airplane.
- Retracts the flaps to the recommended setting; retracts the landing gear if retractable, after a positive rate of climb is established.
- Accelerates to V_x or V_y speed before the final flap retraction; returns to the altitude, heading, and airspeed specified by the examiner.

suggestions and tips for slow flight and stalls practice flights

- These Practice Flights are best suited as demonstration and review tools to help new pilots learn the principles and procedures associated with slow flight and stalls, not to develop specific stick-and-rudder skills.
- Use the views described in Chapter 4, "Flight Simulator Essentials," to provide a better view outside the cockpit. The "virtual cockpit" and "wide" views make it easier it easier to see the visual horizon and to avoid fixating on the instrument panel.

navigation

Use the Practice Flights in this section to observe and practice navigation skills, including:

Pilotage
Dead (or "ded" as in *deduced*) reckoning
Electronic navigation
Flight-planning and maintaining a navigation log
Diversions

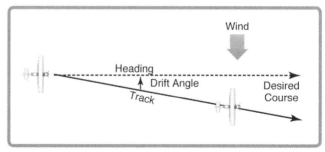

Effects of wind drift on maintaining desired course.
(from FAA-H-8083-25)

background information

Before using the "Navigation" Practice Flights, you should be familiar with the following background information:

Chapter 14: Navigation, in the *Pilot's Handbook of Aeronautical Knowledge*.

"Introduction to VFR Symbols" and "VFR Chart Symbols" in the *Aeronautical Chart User's Guide*.

"What You Need to Know about VOR" and "Using the Radios" in the Flight Simulator Learning Center

The Safety Brief "Terrain Avoidance" available in the free online library at the AOPA Air Safety Foundation website.

performance standards

These Practice Flights apply to Task VII.A "Navigation" and Task VII.B "Navigation Systems and Radar Services" in the FAA Private Pilot PTS, which include the following criteria:

pilotage and dead reckoning

Objective—Determine that the pilot:

- Exhibits knowledge of the elements related to pilotage and dead reckoning.
- Follows the preplanned course by reference to landmarks.
- Identifies landmarks by relating surface features to chart symbols.
- Navigates by means of computed headings, groundspeeds, and elapsed time.
- Corrects for and records the differences between preflight groundspeed and heading calculations and those determined en route.
- Verifies the airplane's position within 3 nautical miles of the flight-planned route.
- Arrives at the enroute checkpoints within 5 minutes of the initial or revised ETA and provides a destination estimate.
- Maintains the appropriate altitude, ±200 feet and headings, ±15°.

navigation systems and radar services

Objective—Determine that the pilot:

- Exhibits knowledge of the elements related to navigation systems and radar services.
- Demonstrates the ability to use an airborne electronic navigation system.
- Locates the airplane's position using the navigation system.
- Intercepts and tracks a given course, radial or bearing, as appropriate.
- Recognizes and describes the indication of station passage, if appropriate.
- Recognizes signal loss and takes appropriate action.
- Maintains the appropriate altitude, ±200 feet and headings, ±15°.

diversions

Objective—Determine that the pilot:

- Exhibits knowledge of the elements related to diversion.
- Selects an appropriate alternate airport and route.
- Makes an accurate estimate of heading, groundspeed, arrival time, and fuel consumption to the alternate airport.
- Maintains the appropriate altitude, ±200 feet (60 meters) and heading, ±15°.

lost procedures

Objective—Determine that the pilot:

- Exhibits knowledge of the elements related to lost procedures.
- Selects an appropriate course of action.
- Maintains an appropriate heading and climbs, if necessary.
- Identifies prominent landmarks.
- Uses navigation systems/facilities and/or contacts an ATC facility for assistance, as appropriate.

suggestions and tips for navigation practice flights

- Use the autopilot at first to learn and practice fundamental navigation skills.
- Use the views described in Chapter 4, "Flight Simulator Essentials," to provide a better view of landmarks. The "virtual cockpit" and "wide" views make it easier to see the visual horizon and to avoid fixating on the instrument panel.
- Keep a flight log, noting waypoints (airports, VORs and VOR cross-bearings, and prominent rivers and bodies of water are the best waypoints in Flight Simulator).

VOR navigation

You can use the VOR Navigation Practice Flights described in Chapter 12, "Introduction to the IFR Practice Flights" to observe and practice navigation using VOR navaids, including the following specific skills:

- Locating information about VOR stations on a chart
- Tuning and verifying VOR signals on the navigation radios
- Determining your position relative to a VOR
- Taking cross-bearings to fix your position and identify intersections
- Intercepting and tracking radials inbound to and outbound from a VOR

NDB navigation

You can use the NDB Navigation Practice Flights described in Chapter 12, "Introduction to the IFR Practice Flights" to observe and practice navigating while using NDB stations, including the following specific skills:

- Determining your position relative to an NDB
- Intercepting and tracking courses inbound to and outbound from an NDB
- Taking cross-bearings to fix your position and identify intersections defined by an NDB bearing and VOR radial or ILS localizer

GPS navigation

These Practice Flights do not apply to a specific task in the Navigation area of operation in the FAA Private Pilot PTS, but they may be useful in helping students and pilots learn basic procedures associated with GPS units.

background information

Before using the GPS Navigation Practice Flights, you should be familiar with the following background information:

Chapter 14: Navigation, in the *Pilot's Handbook of Aeronautical Knowledge*.

The detailed information about the GPS navigation system in AIM 1-1-19.

The topic "Using the GPS" in the Flight Simulator Learning Center, which includes details about the GPS in *Microsoft Flight Simulator*.

The Safety Advisor "GPS Technology" available in the free online library at the AOPA Air Safety Foundation website.

The free online course "GPS for VFR Operations" available at the AOPA Air Safety Foundation website.

The complete Garmin 500 Pilot's Guide available for download from the Garmin website. The Garmin 500 Quick Reference Guide, another handy reference, is also available from the Garmin website. To learn more about GPS, see "About GPS" and "GPS Guide for Beginners" on the Garmin website.

suggestions and tips for GPS navigation practice flights

- Use the autopilot at first to learn and practice fundamental GPS skills.

basic instrument maneuvers

Use the Practice Flights in this section to learn and practice basic attitude instrument flying skills, including:

- Straight-and-level flight at various speeds
- Turns
- Climbs and descents
- Basic instrument flight patterns

Constant power plus constant pitch equals constant airspeed.
(from FAA-H-8083-15)

background information

Before using the Basic Instrument Maneuvers Practice Flights, you should be familiar with the following background information:

Chapter 6: Flight Instruments, *Pilot's Handbook of Aeronautical Knowledge.*

Chapter 3: Flight Instruments, *Instrument Flying Handbook*

Chapter 4: Airplane Attitude Instrument Flying, *Instrument Flying Handbook*

Chapter 5: Airplane Basic Flight Maneuvers, *Instrument Flying Handbook*

"Basic Instrument Flight Patterns" at the end of Chapter 5 of the *Instrument Flying Handbook*

The Safety Advisors "Spatial Disorientation" and "Say Intentions…When you need ATC's help," available in the free online library at the AOPA Air Safety Foundation website.

performance standards

These Practice Flights apply to Task IX "Basic Instrument Maneuvers" in the FAA Private Pilot PTS, which include the following criteria:

straight-and-level flight

Objective—Determine that the pilot:

- Exhibits knowledge of the elements related to attitude instrument flying during straight-and-level flight.
- Maintains straight-and-level flight solely by reference to instruments using proper instrument cross-check and interpretation, and coordinated control application.
- Maintains altitude, ±200 feet (60 meters); heading, ±20°; and airspeed, ±10 knots.

constant airspeed climbs

Objective—Determine that the pilot:

- Exhibits knowledge of the elements related to attitude instrument flying during constant airspeed climbs.
- Establishes the climb configuration specified by the examiner.
- Transitions to the climb pitch attitude and power setting on an assigned heading using proper instrument cross-check and interpretation, and coordinated control application.
- Demonstrates climbs solely by reference to instruments at a constant airspeed to specific altitudes in straight flight and turns.
- Levels off at the assigned altitude and maintains that altitude, ±200 feet (60 meters); maintains heading, ±20°; maintains airspeed, ±10 knots.

constant airspeed descents

Objective—Determine that the pilot:

- Exhibits knowledge of the elements related to attitude instrument flying during constant airspeed descents.
- Establishes the descent configuration specified by the examiner.
- Transitions to the descent pitch attitude and power setting on an assigned heading using proper instrument cross-check and interpretation, and coordinated control application.
- Demonstrates descents solely by reference to instruments at a constant airspeed to specific altitudes in straight flight and turns.
- Levels off at the assigned altitude and maintains that altitude, ±200 feet (60 meters); maintains heading, ±20°; maintains airspeed, ±10 knots.

turns to headings

Objective—Determine that the pilot:

- Exhibits knowledge of the elements related to attitude instrument flying during turns to headings.
- Transitions to the level-turn attitude using proper instrument crosscheck and interpretation, and coordinated control application.
- Demonstrates turns to headings solely by reference to instruments; maintains altitude, ±200 feet (60 meters); maintains a standard rate turn and rolls out on the assigned heading, ± 10°; maintains airspeed, ±10 knots.

recovery from unusual flight attitudes

Objective—Determine that the pilot:

- Exhibits knowledge of the elements related to attitude instrument flying during unusual attitudes.
- Recognizes unusual flight attitudes solely by reference to instruments; recovers promptly to a stabilized level flight attitude using proper instrument cross-check and interpretation and smooth, coordinated control application in the correct sequence.

suggestions and tips for basic instrument maneuvers practice flights

- Use the autopilot at first to observe how the instruments react during basic flight maneuvers skills.
- Use the concept of "the numbers" described in Chapter 8, "Flying the Aircraft Used in the Practice Flights," to establish and maintain stable configurations.
- To enhance these Basic Instrument Maneuvers Practice Flights, use the Basic Attitude Instrument Flying Practice Flights described in Chapter 12, "Introduction to the IFR Practice Flights."

night flying

Use the Practice Flights in this section to learn and practice procedures associated with night flying, including:

- Identifying airports and other features by distinguishing among the types of lights used to aid aerial navigation.
- Using basic instrument flying and navigation skills to enhance aircraft control and navigation.

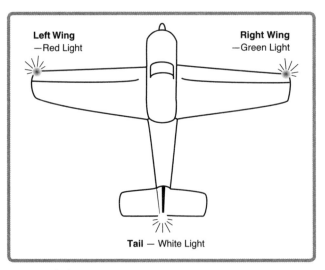

Position lights

background information

Before using the "Night Flying" Practice Flights, you should be familiar with the following background information:

The FARs related to nighttime operations, including: the definition of "night" in Part 1; and sections 61.57, 91.155, 91.151, and 91.209.

AIM Chapter 2, Aeronautical Lighting and Other Airport Visual Aids.

Chapter 10: Night Operations, in the *Airplane Flying Handbook*.

Chapter 15: Aeromedical Factors, in *Pilot's Handbook of Aeronautical Knowledge*.

Chapter 1: Human Factors, in the *Instrument Flying Handbook*.

The Safety Advisors "Spatial Disorientation" and "Say Intentions…When you need ATC's help," available in the free online library at the AOPA Air Safety Foundation website.

The Safety Brief "Terrain Avoidance," available in the free online library at the AOPA Air Safety Foundation website.

performance standards

The FAA Private Pilot PTS do not require demonstration of night-flying skills and procedures. However, you can use the general standards established for instrument flying as a guide.

straight-and-level flight

Objective—Determine that the pilot:

- Exhibits knowledge of the elements related to attitude instrument flying during straight-and-level flight.
- Maintains straight-and-level flight solely by reference to instruments using proper instrument cross-check and interpretation, and coordinated control application.
- Maintains altitude, ±200 feet (60 meters); heading, ±20°; and airspeed, ±10 knots.

suggestions and tips for night flying practice flights

- Use the autopilot at first to reduce workload.
- Use the concept of "the numbers" described in Chapter 8, "Flying the Aircraft Used in the Practice Flights," to establish and maintain stable configurations.
- To enhance these Basic Instrument Maneuvers Practice Flights, use the Basic Attitude Instrument Flying Practice Flights described in Chapter 12, "Introduction to the IFR Practice Flights."

operations in marginal weather

Use the Practice Flights in this section to learn about the difficulties and hazards of attempting to operate under VFR in marginal weather. These Practice Flights may help address such questions as:

- What does "1,000-and-3" weather look like?
- How do the ceilings and visibilities described in FAR §91.155 Basic VFR Weather Minimums apply during a typical flight near a variety of controlled airspace?
- How does marginal weather affect your ability to manage the variety of typical flying tasks, such as navigation, collision avoidance, operating the aircraft systems, and communicating?
- What special hazards are associated with flying in marginal weather in areas with rapidly changing terrain?

background information

Before using the "Operations in Marginal Weather" Practice Flights, you should be familiar with the following background information:

FAR §91.155, Basic VFR Weather Minimums.

Chapter 6: Flight Instruments, *Pilot's Handbook of Aeronautical Knowledge*.

Chapter 3: Flight Instruments, *Instrument Flying Handbook*.

Chapter 4: Airplane Attitude Instrument Flying, *Instrument Flying Handbook*.

Chapter 5: Airplane Basic Flight Maneuvers, *Instrument Flying Handbook*.

Chapter 13: Airspace, *Pilot's Handbook of Aeronautical Knowledge*.

"Introduction to VFR Symbols" and "VFR Chart Symbols" in the *Aeronautical Chart User's Guide.*

The free online course "Know Before You Go: Navigating Today's Airspace," available at the AOPA Air Safety Foundation website.

The Safety Advisors "Spatial Disorientation" and "Say Intentions…When you need ATC's help," "Do the Right Thing—Decision Making for Pilots," "Airspace for Everyone," and "Airspace Flash Cards," available in the free online library at the AOPA Air Safety Foundation website.

The Safety Brief "Terrain Avoidance," available in the free online library at the AOPA Air Safety Foundation website.

You can use "wide view" in Flight Simulator to learn about flying in marginal weather.

suggestions and tips for operations in marginal weather practice flights

- Use the autopilot at first to reduce workload while you observe the out-the-window view in marginal weather.
- Use the views described in Chapter 4, "Flight Simulator Essentials," to provide a better view of horizon. The "virtual cockpit" and "wide" views make it easier to see the visual horizon and to avoid fixating on the instrument panel.
- Use appropriate charts to track position, rehearse communications with ATC facilities, and avoid controlled airspace as required.

understanding airspace

These Practice Flights do not address a specific operational Task in the PTS, but they can help a student or pilot apply conceptual knowledge about airspace to practical situations. The flights begin in the air shortly after departure on a short cross-country flight that takes the aircraft near or through a variety of controlled airspace in the Seattle, WA area. Use the appropriate VFR chart and VOR and DME equipment to fly the route and track position as you fly around and through the airspace. Use the communications radios to simulate contacting the appropriate ATC facilities as required to enter airspace along the route.

These Practice Flights include the following initial conditions:

- Weather is set to restrict your ability to climb above airspace.
- Some Practice Flights may include weather that is below the basic VFR minimums specified in FAR §91.155.

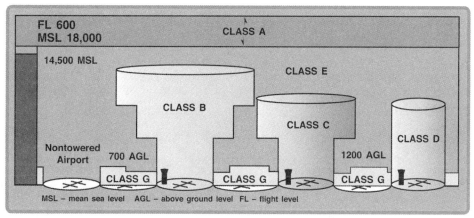

Airspace profile

background information

Before using the "Airspace" Practice Flights, you should be familiar with the following background information:

Chapter 13: Airspace, *Pilot's Handbook of Aeronautical Knowledge*.

AIM Chapter 3, Airspace.

"Introduction to VFR Symbols" and "VFR Chart Symbols" in the *Aeronautical Chart User's Guide*.

The Safety Advisor "Airspace for Everyone" and "Airspace Flash Cards," available in the free online library at the AOPA Air Safety Foundation website.

The free online course "Know Before You Go: Navigating Today's Airspace," available at the AOPA Air Safety Foundation website.

emergencies

Use the Practice Flights in this section to learn and practice procedures associated with some emergencies, including:

- Engine failures during various phases of flight.
- Specific systems and instrument failures.

As suggested in Chapter 3, "Best Practices for Using Flight Simulator," these Practice Flights can help an instructor assess a student's practical knowledge of how to deal with typical problems. The flights can also enhance discussions and demonstrations of such topics as how changes in airspeed and configuration affect an aircraft's power-off descent rate, and how much time, altitude, and maneuvering room is required to return to the runway after an engine failure.

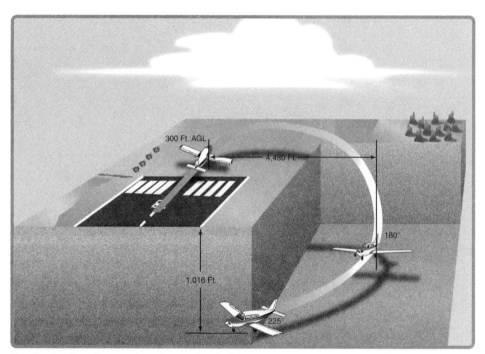

Turning back to the runway after engine failure.

background information

Chapter 5: Aircraft Systems, *Pilot's Handbook of Aeronautical Knowledge*
Chapter 16: Emergency Procedures, *Airplane Flying Handbook*.

performance standards

These Practice Flights apply to Task X "Emergency Operations" in the FAA Private Pilot PTS, which include the following criteria:

emergency approach and landing (simulated)

Objective—Determine that the pilot:

- Exhibits knowledge of the elements related to emergency approach and landing procedures.
- Analyzes the situation and selects an appropriate course of action.
- Establishes and maintains the recommended best-glide airspeed, ±10 knots.

- Selects a suitable landing area.
- Plans and follows a flight pattern to the selected landing area considering altitude, wind, terrain, and obstructions.
- Prepares for landing, or go-around, as specified by the examiner.
- Follows the appropriate checklist.

systems and equipment malfunctions

1. Exhibits knowledge of the elements related to system and equipment malfunctions appropriate to the airplane provided for the practical test.
2. Analyzes the situation and takes appropriate action for simulated emergencies appropriate to the airplane provided for the practical test for at least three (3) of the following—

   - Partial or complete power loss.
   - Engine roughness or overheat.
   - Loss of oil pressure.
   - Fuel starvation.
   - Electrical malfunction.
   - Vacuum/pressure, and associated flight instruments malfunction.
   - Pitot/static.
   - Landing gear or flap malfunction.
   - Inoperative trim.
   - Structural icing.
   - Smoke/fire/engine compartment fire.
   - Any other emergency appropriate to the airplane.

3. Follows the appropriate checklist or procedure.

commercial pilot practice flights

A typical commercial pilot training syllabus emphasizes operations in complex aircraft and smooth, precise use of the aircraft controls through a series of performance and ground reference maneuvers. PC-based flying is not well suited to developing the latter skills, but it is especially useful as a tool for introducing the operating principles of complex aircraft. The Baron makes a good platform for learning proper sequences for setting power, operating cowl flaps, landing gear, and flaps during typical phases of flight.

To complement lessons based on many areas of operation in the Commercial Pilot PTS, you can use the comparable Private Pilot Practice Flights and impose tighter standards (e.g., altitudes within 100 feet, airspeeds within 5 knots) and adjust the procedures for specific maneuvers such as stalls as specified in the Commercial Pilot PTS. Using the Private Pilot Practice Flights with the Baron also supports development of such skills as:

- Rehearsing operations of complex aircraft in the traffic pattern.
- Becoming accustomed to greater rates of climb, higher cruise speeds, and descents at higher speeds.

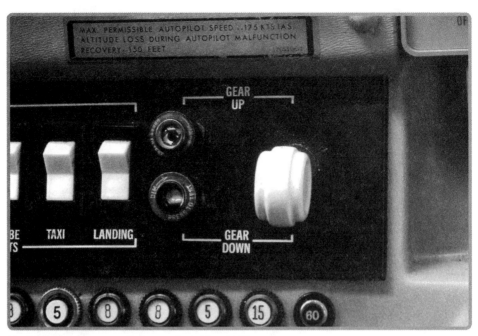

Typical landing gear switches and position indicators.

background information

The references in each category of Private Pilot Practice Flights also apply to the Commercial Pilot Practice Flights. The following additional references are also helpful:

The sections about propellers in Chapter 3: Aerodynamics of Flight, in the *Pilot's Handbook of Aeronautical Knowledge*

Chapter 5: Aircraft Systems, *Pilot's Handbook of Aeronautical Knowledge*

Chapter 11: Transition to Complex Airplanes, *Airplane Flying Handbook*

eleven

sample briefings for VFR practice flights

checklist

- ☐ basic flying skills
- ☐ airport operations
- ☐ background information

This chapter includes examples of the preflight briefings that accompany the VFR Practice Flights. As explained in Chapter 10, "Introduction to the VFR Practice Flights," each Practice Flight includes a detailed briefing in PDF format that you can print to review. The complete set of preflight briefings is available on the CD.

location

Unless noted otherwise, most VFR Practice Flights begin in the airspace around Seattle, WA.

basic flying skills

These Practice Flights support learning and rehearsing basic skills, such as the:

- Purpose and effect of the primary flight and engine controls.
- Purpose and function of the basic flight instruments.
- Operation of the engine controls as well as purpose and function of the engine instruments.

- Use of elevator trim.
- Configuring the aircraft for various situations by applying the Power + Configuration/Attitude = Performance equation as described in Chapter 8, "Flying the Aircraft Used in the Practice Flights."
- Correlating information from instruments and visual flying according to the principle of Integrated Flight Instruction
- You can use the Baron version of this Practice Flight to introduce operating an aircraft with a constant-speed propeller, retractable landing gear, and advanced instruments like the HSI.

These Practice Flights begin in the air southwest of Seattle. You are at 4500 ft in the Cessna 172 Skyhawk or the Beechcraft BE58 Baron with the default instrument panel. The autopilot is ON in HDG and ALT modes, and the simulation is paused.

practice flight file name

flight file name	conditions
BruceAir-VFR-BasicFlyingSkills-C172-01.Flt	Day/VMC/Lt wind
BruceAir-VFR-BasicFlyingSkills-BE58-01.Flt	Day/VMC/Lt wind

background information

Before using the "Basic Flying Skills" Practice Flights, you should be familiar with the following background information:

Chapter 4: Flight Controls, in the *Pilot's Handbook of Aeronautical Knowledge*.

Chapter 6: Flight Instruments, *Pilot's Handbook of Aeronautical Knowledge*.

Chapter 3: Aerodynamics of Flight, *Pilot's Handbook of Aeronautical Knowledge*.

Chapter 3: Basic Flight Maneuvers, *Airplane Flying Handbook*.

The sections about reciprocating engines in Chapter 5, Aircraft Systems, from the *Pilot's Handbook of Aeronautical Knowledge*

starting position on Flight Simulator map

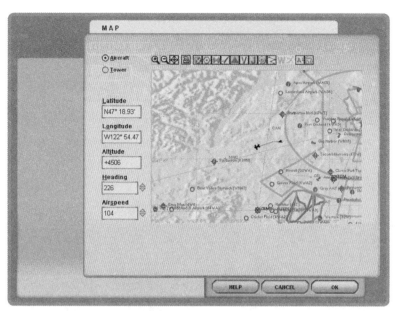

After loading a Practice Flight, you can confirm your starting position on the map in Flight Simulator.

initial instrument panel configurations

Most VFR Practice Flights begin in the air, with the default instrument panel in view.

sample
briefings
for VFR
practice
flights

181

Some of the VFR Practice Flights use the Beechcraft BE58 Baron to support lessons that involve complex, high-performance aircraft.

airport operations—traffic patterns

The Practice Flights in this section support learning about operations in the airport traffic pattern, including:

- Gathering information about airports from a sectional chart and applying that information to entering and flying traffic patterns at towered and non-towered airports.
- Orientation of the aircraft with respect to standard, non-standard, and abbreviated traffic patterns.
- Configuring the aircraft for various situations by applying the Power + Configuration/Attitude = Performance equation as described in Chapter 8, "Flying the Aircraft Used in the Practice Flights."
- Integrating cockpit flow checks/checklists with operations in the vicinity of an airport.
- Setting up the communications radios to receive ATIS and ASOS reports and to communicate with a control tower or on a CTAF, as appropriate

background information

Before using the "Airport Operations—Traffic Patterns" Practice Flights, you should be familiar with the following background information:

Chapter 12: Airport Operations, in the *Pilot's Handbook of Aeronautical Knowledge*.

Chapter 7: Airport Traffic Patterns, in the *Airplane Flying Handbook*.

The Safety Advisors "Operations at Towered Airports" and "Operations at Nontowered Airports" available in the free online library at the AOPA Air Safety Foundation website.

Advisory Circulars 90-42F "Traffic Advisory Practices at Airports Without Operating Control Towers" and 90-66A "Recommended Standard Traffic Patterns and Practices for Aeronautical Operations at Airports without Operating Control Towers."

practice flight file names

flight file name	conditions
BruceAir-VFR-KSHN-TrafficPattern-C172-01.Flt	Day/VMC/Lt wind
BruceAir-VFR-KSHN-TrafficPattern-BE58-01.Flt	Day/IMC/Wind

location

These Practice Flights begin in the air southwest of the SEA VOR, near the KSHN, KPWT, KTIW, and KOLM airports.

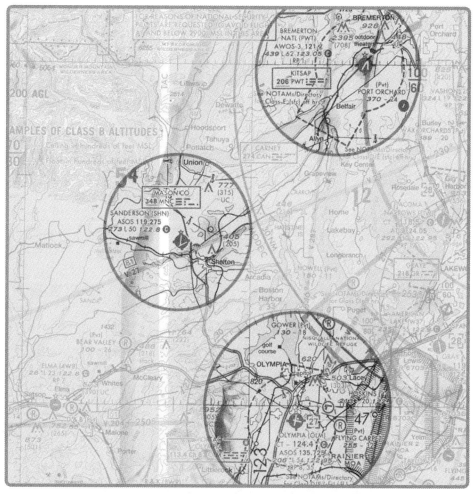

The briefings for the Practice Flights include links to VFR charts you can review before you start flying.

sample
briefings
for VFR
practice
flights

183

starting position as shown on Flight Simulator map

You can use the map in Flight Simulator to compare the starting position for a Practice Flight with the indications on the navigation instruments.

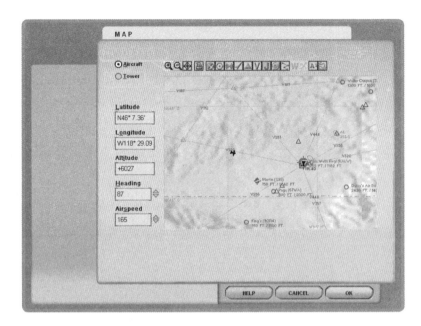

initial instrument panel configurations

The Practice Flights begin with aircraft in a stable configuration, with the navigation radios tuned to appropriate VORs and other navigation aids.

Practice Flights that use the Beechcraft BE58 Baron usually start at the same locations as Flights that use the C172.

sample
briefings
for VFR
practice
flights

185

twelve

introduction to the IFR practice flights

checklist

☐ information, categories, and briefings for IFR practice flights

☐ basic instrument flying

☐ VOR and NDB

☐ departure, en route, and arrivals

☐ holding

☐ procedure turns

☐ DME arcs

☐ approaches— localizer, VOR, NDB, GPS

☐ missed and circling approaches

This chapter describes the categories of IFR Practice Flights available on the CD and from the website for this book. The complete, current list of IFR Practice Flights is available on the website for this book. For background information about the Flights feature in *Microsoft Flight Simulator* and how to use the Practice Flights, see Chapter 7, "About the Practice Flights."

The explanation of each category of IFR Practice Flights includes:

- The purpose of the Practice Flights included in the category.
- References to the background information from FAA training handbooks, the Aeronautical Information Manual, and other resources that you should review before flying Practice Flights in that category.
- General information about the locations used for the Practice Flights.
- Excerpts from the Performance Standards from the relevant FAA Practical Test Standards to help you evaluate your flights. Note that the Performance Standards listed for each Practice Flight do not include all of the standards and evaluation criteria from the official test standards. For the complete set of applicable criteria, refer to the latest official versions of the Practical Test Standards.
- Suggestions and tips for how to use the Practice Flights in that category most effectively, whether you're a student, pilot, or flight instructor.

The preflight briefing included with each Practice Flight (in Adobe Reader format) provides additional information, including:

- The type of aircraft used in each variation of that Practice Flight.
- The starting location of that Practice Flight.
- The weather, time of day, and other pertinent information about the environment for that Practice Flight.

- Any specific instructions, including, when appropriate, an ATC clearance for that Practice Flight.
- A chart or chart excerpt that covers the area in which that Practice Flight takes place or the instrument approach or departure procedure associated with that Practice Flight. The IFR charts are reproduced from the Voyager Flight Software System with the permission of Seattle Avionics Software, Inc. and obviously should be used only for simulated flying, not for navigation.

important note

These IFR Practice Flights reflect the procedures that were in effect when I created the flights. Some of the information they are based on may be out of date, and the database in *Microsoft Flight Simulator* is not revised to include the additions, corrections, and deletions incorporated in databases required for real flight. You may find that some procedures, navaid frequencies, altitudes, and other details are no longer valid or correct. In particular, note that GPS procedures published after *Microsoft Flight Simulator* was released are not included in the Flight Simulator database for the GPS.

Obviously, you must use current charts and other information when you fly specific IFR approaches in the real world. In other words, these Practice Flights are useful as training aids to help you understand and fly *representative* types of IFR procedures, but they are not intended to train or prepare you to fly specific routes, approaches or departures. For more information about these issues, see "The Flight Simulator Database and IFR Procedures" in Chapter 2, "Using Flight Simulator as a Training Aid."

sources of background information

Excerpts from public-domain resources such as the FAA training handbooks are provided in PDF format on the CD and on the website for this book. For detailed descriptions of the resources, see Chapter 1, "About this Book" and Chapter 9, "Supplemental Information."

categories of IFR practice flights

I have organized the IFR Practice Flights into categories and listed them here in a sequence that follows a typical IFR training syllabus. (The *Instrument Flying Handbook* includes a sample IFR syllabus in Appendix 2.) Of course, you can use the Practice Flights in any order you choose, but if you're a newcomer to IFR flying, following this sequence will help you develop essential basic skills that prepare you to handle advanced procedures.

The IFR Practice Flights are collected into the following basic categories, each described in detail later in this chapter:

- Basic Attitude Instrument Flying (BAI)
- VOR Navigation
- NDB (ADF) Navigation
- Holding Procedures
- Departure Procedures
- Procedure Turns
- En Route Navigation and Procedures
- Arrivals (STARs)
- Holding Patterns
- DME Arcs
- Approaches
- Instrument Landing System (ILS)
- Localizer (LOC, BC, LDA)

- VOR (VOR and VOR-DME)
- NDB
- GPS
- Missed Approachs
- Circling Approaches

sample preflight briefings for IFR practice flights

You can find examples of typical briefings for IFR Practice Flights in Chapter 13, "Briefings for IFR Practice Flights."

basic attitude instrument flying (BAI)

Use the Practice Flights in this section to observe and practice fundamental instrument flying skills, including:

- Identifying and understanding the functions of the flight and navigation instruments
- Developing an effective instrument scan
- Applying the Control–Performance and Primary-Supporting methods of aircraft control to basic attitude instrument flying
- Developing your understanding of how to perform basic flight maneuvers, such as standard-rate turns, and constant-airspeed and constant-rate climbs and descents.
- Flying the Basic Instrument Flight Patterns that are components of published instrument procedures, including:
  –Racetrack Pattern
  –Standard Procedure Turn
  –80/260 Procedure Turn
  –Teardrop Pattern
  –Circling Approaches Patterns I and II

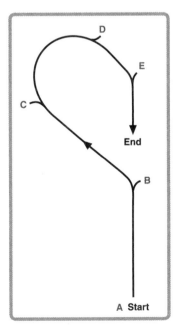

Standard procedure turn
(entire pattern in level flight)

background information

Before using the BAI Practice Flights, you should be familiar with the following background information:

Chapter 3: Flight Instruments, in the *Instrument Flying Handbook*.

Chapter 4: Airplane Attitude Instrument Flying, *Instrument Flying Handbook*.

Chapter 5: Airplane Basic Flight Maneuvers, *Instrument Flying Handbook*.

"Basic Instrument Flight Patterns," at the end of Chapter 5 in the *Instrument Flying Handbook*.

locations for BAI practice flights

The BAI Practice Flights begin in the airspace southwest of Seattle, WA. You can use ground-based navaids in the area for general orientation, but none of the BAI Practice Flights specifically includes tracking a course (i.e., a flight path defined by a VOR radial or track to a GPS waypoint).

performance standards

These Practice Flights apply to Task IV. A., "Flight by Reference to Instruments," in the FAA IFR Practical Test Standards, which includes the following basic standards for flying solely by reference to instruments.

Objective—To determine that the applicant:

- Exhibits adequate knowledge of the elements related to attitude instrument flying during straight-and-level, climbs, turns, and descents while conducting various instrument flight procedures.
- Maintains altitude within +/– 100 feet during level flight, headings within +/– 10°, airspeed within +/– 10 knots, and bank angles within +/– 5° during turns.
- Uses proper instrument cross-check and interpretation, and applies the appropriate pitch, bank, power, and trim corrections when applicable.

suggestions and tips for BAI practice flights

- Use the concept of "the numbers" described in Chapter 8, "Flying the Aircraft Used in the Practice Flights," to help you establish and maintain stable configurations.
- Associate specific pitch and bank attitudes, power settings, and aircraft configurations with aircraft performance.
- Use the autopilot at first so that you can observe how the instruments react as the airplane turns, climbs, and descends.

VOR navigation

Use the Practice Flights in this section to observe and practice navigation using VOR navaids, including the following specific skills:

- Locating information about VOR stations on a chart.
- Tuning and verifying VOR signals on the navigation radios.
- Determining your position relative to a VOR.
- Taking cross-bearings to fix your position and identify intersections.
- Intercepting and tracking radials inbound to and outbound from a VOR.

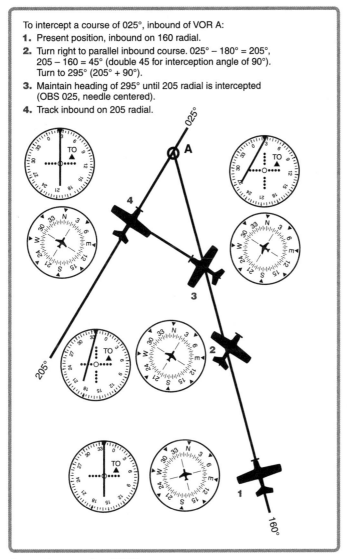

To intercept a course of 025°, inbound of VOR A:

1. Present position, inbound on 160 radial.
2. Turn right to parallel inbound course. 025° − 180° = 205°, 205 − 160 = 45° (double 45 for interception angle of 90°). Turn to 295° (205° + 90°).
3. Maintain heading of 295° until 205 radial is intercepted (OBS 025, needle centered).
4. Track inbound on 205 radial.

Course interception (VOR)

Note that you can also use these VOR Navigation Practice Flights to learn about and practice using HSI and RMI instruments. Until recently, these instruments were installed primarily in high-performance aircraft, but with the development of "glass cockpits," both the HSI and RMI (bearing-pointer) presentations are appearing even in trainers and basic personal aircraft. To practice using these instruments, fly either the Cessna 172SP or Baron equipped with the Garmin G1000 "glass cockpit" or the Baron with the standard instrument panel.

You can use the Practice Flights for the Beechcraft BE58 Baron to practice using an HSI and RMI.

background information

Before using the VOR Navigation Practice Flights, you should be familiar with the following background information:

> The sections about VOR navigation in Chapter 7 of the *Instrument Flying Handbook*.

> The information about DME in Chapter 7 of the *Instrument Flying Handbook*.

> The information about the VOR navigation system in AIM 1-1-3.

> "What You Need to Know about VOR" and "Using the Radios" in the Flight Simulator Learning Center.

Additional background information about VOR and DME is available in Chapter 14 of the *Pilot's Handbook of Aeronautical Knowledge*.

locations for VOR navigation IFR practice flights

The VOR Navigation Practice Flights take place the airspace around Walla Walla, Washington.

performance standards

These Flights apply to Area of Operation V, "Navigation Systems" in the FAA IFR Practical Test Standards, which includes the following performance standards:

Objective—To determine that the applicant:

- Exhibits adequate knowledge of the elements related to intercepting and tracking navigational systems.
- Tunes and correctly identifies the navigation facility.
- Sets and correctly orients the course to be intercepted into the course selector or correctly identifies the course on the RMI.

- Intercepts the specified course at a predetermined angle, inbound or outbound from a navigational facility.
- Maintains the airspeed within +/–10 knots, altitude within +/–100 feet, and selected headings within +/–5°.
- Applies proper correction to maintain a course, allowing no more than three-quarter-scale deflection of the CDI or within +/–10° in case of an RMI.
- Determines the aircraft position relative to the navigational facility.
- Recognizes navigational receiver or facility failure, and when required, reports the failure to ATC.

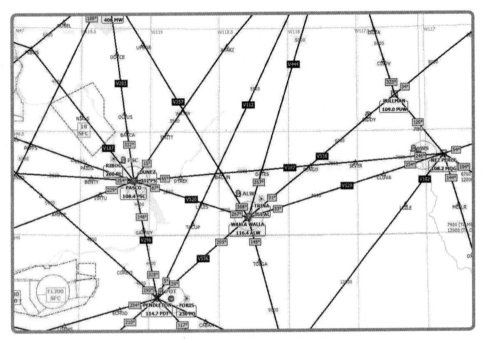

The VOR navigation flights begin the air near Walla Walla, WA (ALW)

NDB navigation

Use the Practice Flights in this section to observe and practice navigating while using NDB stations, including the following specific skills:

- Determining position relative to an NDB
- Intercepting and tracking courses inbound to and outbound from an NDB
- Taking cross-bearings to fix your position and identify intersections defined by an NDB bearing and VOR radial or ILS localizer

Note that you can also use these NDB Navigation Practice Flights to learn about and practice using the RMI instrument in the Baron equipped with a standard instrument panel.

background information

Before using the NDB Navigation Practice Flights, you should be familiar with the following background information:

The sections about NDB navigation in Chapter 7 of the *Instrument Flying Handbook*.

The detailed information about the NDB/ADF navigation system in AIM 1-1-2.

"Automatic Direction Finder" and "Using the Radios" in the Flight Simulator Learning Center.

locations for the NDB navigation practice flights

The NDB Practice Flights take place in the airspace near Shelton, WA (KSHN) and use the Mason County (MNC) and Bremerton, WA (PWT) NDBs.

performance standards

These Flights apply to Area of Operation V, "Navigation Systems" in the FAA IFR Practical Test Standards. The criteria outlined in the Performance Standards section for the VOR Navigation Practice Flights also apply to these Flights.

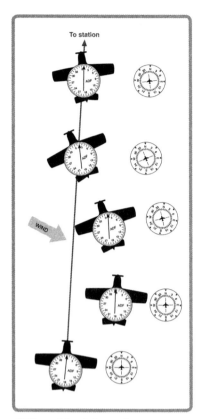

ADF tracking inbound

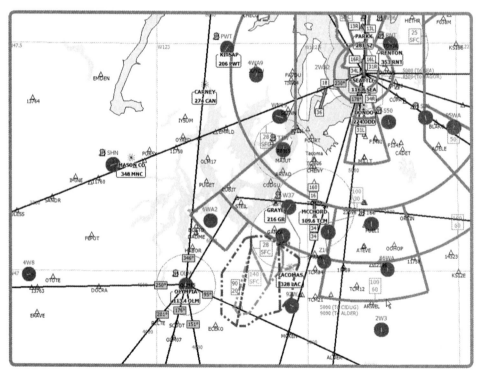

Starting area for NDB navigation Practice Flights

departure procedures

Use the Practice Flights in this section to observe and practice flying published IFR Standard Instrument Departure procedures (SIDs) and related skills, including:

- Flying a SID
- Flying an Obstacle Departure Procedure (ODP) from an airport without a charted departure procedure

background information

Before using the Departure Procedures Practice Flights, you should be familiar with the following background information:

The sections about Departure Procedures in Chapter 10 of the *Instrument Flying Handbook*.

Chapter 2 of the *Instrument Procedures Handbook*.

The detailed information about departure procedures in AIM 5-2-5.

locations for the departure procedures practice flights

The Departure Procedures Practice Flights use SIDs and departure procedures at the Rogue Valley International Airport at Medford, OR (KMFR) and other airports in the Pacific Northwest.

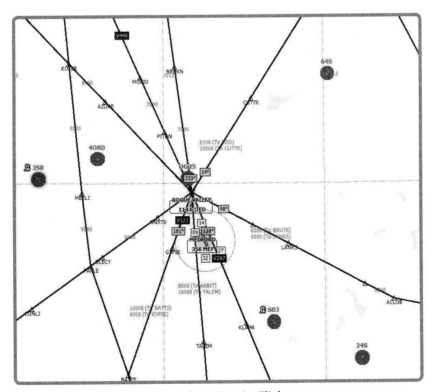

Starting area for departure procedures Practice Flights

performance standards

These Practice Flights relate to Area of Operation III, "Air Traffic Control Clearances and Procedures" in the FAA IFR Practical Test Standards.

The applicable objectives include determining that the pilot:

- Selects and uses the appropriate communication facilities; selects and identifies the navigation aids associated with the proposed flight.
- Performs the appropriate aircraft checklist items relative to the phase of flight.
- Exhibits adequate knowledge of communication failure procedures.
- Intercepts, in a timely manner, all courses, radials, and bearings appropriate to the procedure, route, or clearance.
- Maintains the applicable airspeed within +/–10 knots; headings within +/–10°; altitude within +/–100 feet; and tracks a course, radial or bearing within ¾ scale deflection of the CDI.

en route navigation and procedures

Use the Practice Flights in this section to observe and practice en route IFR flying skills, including:

- Joining the en route airway structure after departure.
- Identifying intersections, VOR changeover points, and minimum en route altitudes along an airway.
- Maintaining the proper en route altitudes for each section of the cleared route.
- Changing course at VORs and intersections.
- Using GPS for en route operations.
- Planning descents.

background information

The information about airspace and en route operations in Chapter 8 of the *Instrument Flying Handbook*.

The sections about en route operations in Chapter 3 of the *Instrument Procedures Handbook*.

AIM Section 3, En Route Procedures.

The Safety Advisors "Single-Pilot IFR" and "Weather Wise" available in the online library at the AOPA Air Safety Foundation website.

locations for the en route navigation and procedures flights

The En Route Navigation and Procedures Practice Flights take place in the airspace near Seattle, WA (SEA) (see Page 194), and Wenatchee, WA (EAT).

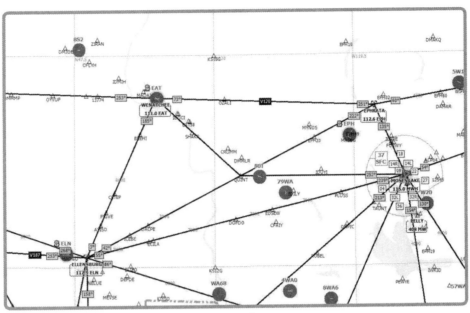

Starting area for en route navigation and procedures

performance standards

These Practice Flights relate to Area of Operation V, "Navigation Systems" in the FAA IFR Practical Test Standards, which includes the following standards:

Objective—To determine that the applicant:

- Tunes and correctly identifies the navigation facility.
- Sets and correctly orients the course to be intercepted into the course selector or correctly identifies the course on the RMI.
- Intercepts the specified course at a predetermined angle, inbound or outbound from a navigational facility.
- Maintains the airspeed within +/–10 knots, altitude within +/–100 feet, and selected headings within +/–5°.
- Applies proper correction to maintain a course, allowing no more than three-quarter-scale deflection of the CDI or within +/–10° in case of an RMI.
- Determines the aircraft position relative to the navigational facility or from a waypoint in the case of GPS.
- Recognizes navigational receiver or facility failure, and when required, reports the failure to ATC.

arrivals (STARs) practice flights

Use the Practice Flights in this section to observe and practice Standard Terminal Area Arrival Routes (STARs).

background information

Before using the Arrival Procedures Practice Flights, you should be familiar with the following background information:

The sections about arrival procedures in Chapter 7 of the *Instrument Flying Handbook*.

The information about arrivals in Chapter 4 of the *Instrument Procedures Handbook*.

The sections about holding procedures in Chapter 3 of the *Instrument Procedures Handbook.*

Basic hold terminology as illustrated in AIM Figure 5-3-3.

Types of holds as illustrated in AIM Figure 5-3-2.

The detailed information about Holding in AIM 5-3-7.

locations for arrival procedures practice flights

The Arrival Procedures Practice Flights use the airspace around Seattle, WA, Spokane, WA, and Portland, OR, which include STARs for piston-powered light aircraft.

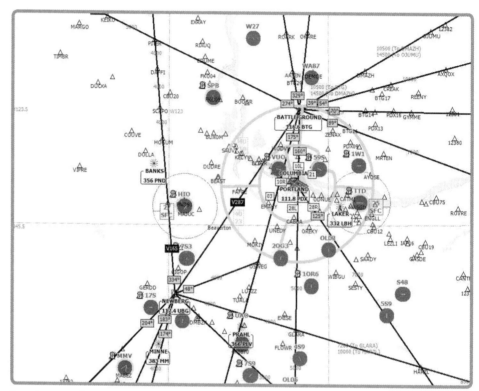

Starting area for arrival procedures

performance standards

These Practice Flights relate to Area of Operation III B, "Compliance with Departure, En Route, and Arrival Procedures and Clearances" in the FAA IFR Practical Test Standards, which includes the following standards:

Objective. To determine that the applicant:

- Exhibits adequate knowledge of the elements related to ATC routes, and related pilot/controller responsibilities.
- Uses the current and appropriate navigation publications for the proposed flight.
- Selects and uses the appropriate communication facilities and selects and identifies the navigation aids associated with the proposed flight.
- Performs the appropriate aircraft checklist items relative to the phase of flight.
- Establishes two-way communications with the proper controlling agency, using proper phraseology.
- Complies, in a timely manner, with all ATC instructions and airspace restrictions.

holding procedures

Use the Practice Flights in this section to observe and practice holding procedures, including holds:

- At a VOR
- At an NDB or LOM
- At an intersection defined by two VOR radials
- On a localizer
- At a DME fix on a VOR radial

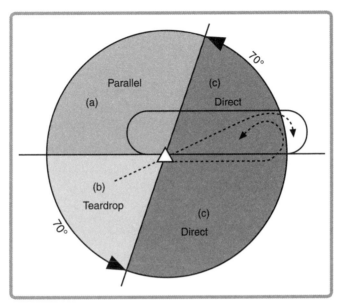

Holding pattern entry procedures

background information

Before using the Holding Procedures Practice Flights, you should be familiar with the following background information:

The sections about Holding Procedures in Chapter 7 of the *Instrument Flying Handbook*.

The sections about Holding Procedures in Chapter 3 of the *Instrument Procedures Handbook*.

Basic hold terminology as illustrated in AIM Figure 5-3-3.

Types of holds as illustrated in AIM Figure 5-3-2.

The detailed information about Holding in AIM 5-3-7.

locations for holding procedures practice flights

The Holding Procedures Practice Flights use the airspace around Seattle, WA. Several of the Flights use navaids and intersections southwest of the Seattle VOR (SEA).

performance standards

These Flights apply to Area of Operation III.C, "Holding Procedures" in the FAA IFR Practical Test Standards, which includes the following performance standards:

Objective—To determine that the applicant:

- Exhibits adequate knowledge of the elements related to holding procedures.
- Changes to the holding airspeed appropriate for the altitude or aircraft when 3 minutes or less from, but prior to arriving at, the holding fix.
- Explains and uses an entry procedure that ensures the aircraft remains within the holding pattern airspace for a standard, nonstandard, published, or non-published holding pattern.
- Recognizes arrival at the holding fix and initiates prompt entry into the holding pattern.
- Complies with ATC reporting requirements.
- Uses the proper timing criteria, where applicable, as required by altitude or ATC instructions.
- Complies with pattern leg lengths when a DME distance is specified.
- Uses proper wind correction procedures to maintain the desired pattern and to arrive over the fix as close as possible to a specified time.
- Maintains the airspeed within +/−10 knots; altitude within +/−100 feet; headings within +/−10°; and tracks a selected course, radial or bearing within ¾ scale deflection of the CDI.

VOR holds

The VOR holding procedures focus on the Olympia, WA (OLM) VOR. Some of the Flights include holding instructions in an initial clearance. Others use the charted (non-standard) hold at OLM.

intersection holds

Intersection holds are based on the CARRO fix, defined by the intersection of the SEA 230 radial and the OLM 346 radial, 19 nm north of OLM.

The Flights, which begin at different starting points, include holding instructions in an initial clearance.

NDB holds

The NDB holding procedures use the MNC beacon near the Shelton, WA (KSHN) airport, about 36 nm west-southwest of Seattle.

Some of the Flights use the charted (non-standard) hold that is part of the NDB GPS-A approach at KSHN. Other Flights include a variety of holds as issued in an initial clearance.

localizer holds

The localizer holding Practice Flights use the hold charted on the ILS RWY 14 approach at the Medford, OR (KMFR) airport, and other airports in the Pacific Northwest.

DME fix holds

The holding procedures Practice Flights for DME holds use the CARRO intersection located on V27 24 nm southwest of Seattle. You can also practice DME holds associated with IAPs used in other Practice Flights.

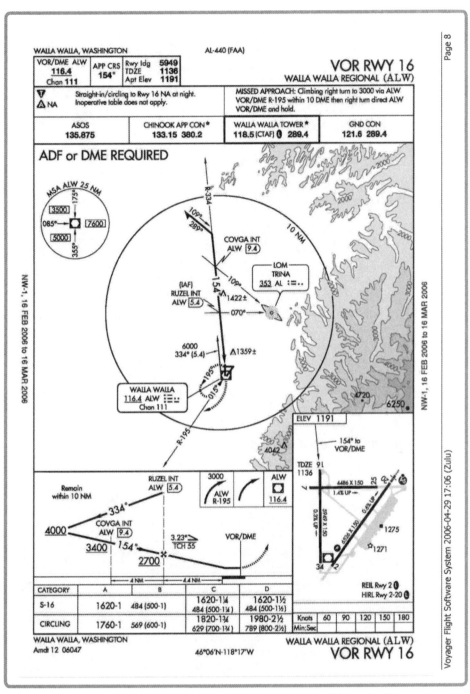

You practice holding procedures by using any of the IFR Practice Flights for IAPs, such as this VOR approach at Walla Walla, WA.

procedure turns

Use the Practice Flights in this section to observe and practice course reversals used in many instrument approach procedures, including:

- Standard Procedure Turn
- 80/260 Procedure Turn
- Teardrop Pattern

background information

Before using the Procedure Turns Practice Flights, you should be familiar with the following background information:

The sections about procedure turns in Chapter 8 of the *Instrument Flying Handbook*.

The information about procedure turns in Chapter 5 of the *Instrument Procedures Handbook*.

The detailed information about procedure turns in AIM 5-4-9.

"Basic Instrument Flight Patterns" at the end of Chapter 5 of the *Instrument Flying Handbook*.

locations for procedure turns practice flights

The Procedure Turns Practice Flights use approach procedures from other categories of IFR Practice Flights, such as the VOR Rwy 16 approach at KALW.

performance standards

These Practice Flights relate to Area of Operation VI, "Instrument Approach Procedures" in the FAA IFR Practical Test Standards. Task A, "Nonprecision Approach," requires that an applicant fly two nonprecision approaches, at least one of which must include a procedure turn.

The applicable standards include determining that the pilot:

- Exhibits adequate knowledge of the elements related to an instrument approach procedure.
- Selects and complies with the appropriate instrument approach procedure to be performed.
- Selects, tunes, identifies, and confirms the operational status of navigation equipment to be used for the approach procedure.
- Complies with all clearances issued by ATC or the examiner.
- Recognizes if any flight instrumentation is inaccurate or inoperative, and takes appropriate action.
- Advises ATC or examiner anytime that the aircraft is unable to comply with a clearance.
- Establishes the appropriate aircraft configuration and airspeed considering turbulence and wind shear, and completes the aircraft checklist items appropriate to the phase of the flight.
- Maintains, prior to beginning the final approach segment, altitude within +/–100 feet, heading within +/–10° and allows less than ¾ scale deflection of the CDI or within +/–10° in the case of an RMI, and maintains airspeed within +/–10 knots.

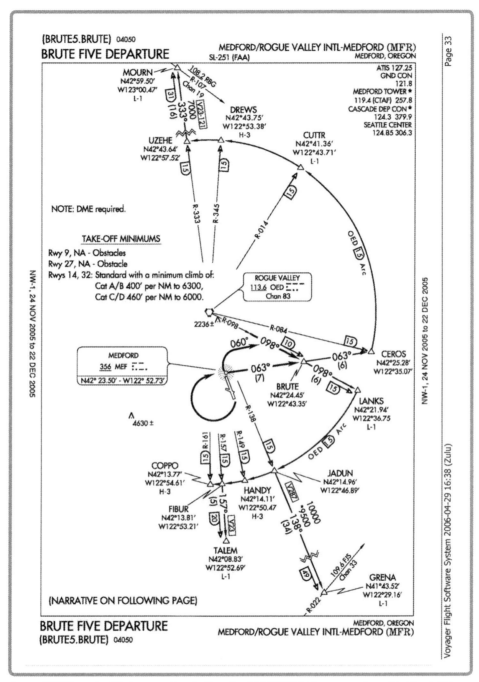

(BRUTE5.BRUTE) 04050
BRUTE FIVE DEPARTURE
SL-251 (FAA)

MEDFORD/ROGUE VALLEY INTL-MEDFORD (MFR)
MEDFORD, OREGON

ATIS 127.25
GND CON
121.8
MEDFORD TOWER *
119.4 (CTAF) 257.8
CASCADE DEP CON *
124.3 379.9
SEATTLE CENTER
124.85 306.3

MOURN
N42°59.50'
W123°00.47'
L-1

108.2 RBG
R-107
Chan 19

333°
(16)
7000

DREWS
N42°43.75'
W122°53.38'
H-3

CUTTR
N42°41.36'
W122°43.71'
L-1

UZEHE
N42°43.64'
W122°57.52'

R-333
R-345

15
15

OED 15 Arc

NOTE: DME required.

TAKE-OFF MINIMUMS
Rwy 9, NA - Obstacles
Rwy 27, NA - Obstacle
Rwys 14, 32: Standard with a minimum climb of:
    Cat A/B 400' per NM to 6300,
    Cat C/D 460' per NM to 6000.

ROGUE VALLEY
113.6 OED ⋅ ⋅ ⋅
Chan 83

2236± R-098 R-084

15

060° 098° 10
063°
098° (6)
CEROS
N42°25.28'
W122°35.07'

MEDFORD
356 MEF ⋅ ⋅ ⋅
N42° 23.50' - W122° 52.73'

063°
(7)
BRUTE
N42°24.45'
W122°43.35'

098°
(6)

15

LANKS
N42°21.94'
W122°36.75
L-1

Λ
4630 ±

R-138

OED 15 Arc

R-161
R-157
R-149

15 15 15

15

COPPO
N42°13.77'
W122°54.61'
H-3

JADUN
N42°14.96'
W122°46.89'

HANDY
N42°14.11'
W122°50.47
H-3

V287

10000
•9500
138°
(34)

FIBUR
N42°13.81'
W122°53.21'

157°
(5)

20

V21

TALEM
N42°08.83'
W122°52.69'
L-1

109.6 EIS
Chan 33

49

GRENA
N41°43.52'
W122°29.16'
L-1

R-022

(NARRATIVE ON FOLLOWING PAGE)

BRUTE FIVE DEPARTURE
(BRUTE5.BRUTE) 04050

MEDFORD, OREGON
MEDFORD/ROGUE VALLEY INTL-MEDFORD (MFR)

NW-1, 24 NOV 2005 to 22 DEC 2005

NW-1, 24 NOV 2005 to 22 DEC 2005

Voyager Flight Software System 2006-04-29 16:38 (Zulu)

Several IFR Practice Flights use the challenging BRUTE FIVE departure from Medford, OR, which includes DME arcs.

DME arcs

Use the Practice Flights in this section to observe and practice flying DME arcs used in many instrument procedures, including:

- Intercepting DME arcs as part of a charted transition from the en route airway structure to an arrival or to an IAF, or as part of a departure procedure to the en route structure.
- Intercepting DME arcs from inside and outside the arc
- Maintaining a DME arc with wind blowing you inside or outside the arc
- Flying DME arcs with VOR, HSI, and RMI indicators
- Using lead radials

background information

Before using the DME Arcs Practice Flights, you should be familiar with the following background information:

The sections about VOR in Chapter 7 of the *Instrument Flying Handbook*.

The information about DME in Chapter 7 of the *Instrument Flying Handbook*.

The information about DME arcs in Chapter 5 of the *Instrument Procedures Handbook*.

The information about the VOR in AIM 1-1-3.

The information about DME in AIM 1-1-7.

To fly DME arcs while using GPS for course guidance, you must also be familiar with the following information:

- "Use of GPS in lieu of ADF and DME" in AIM 1-1-19(f)
- The topic "Using the GPS" in the Flight Simulator Learning Center, which includes details about the GPS in *Microsoft Flight Simulator*
- The specific procedures for flying DME arcs with the Garmin 500-series GPS in *Microsoft Flight Simulator*. You can download the complete Garmin 500 Pilot's Guide from the Garmin website. The Garmin 500 Quick Reference Guide, another handy reference, is also available from the Garmin website.

Other useful sources of information about GPS include:

- The AOPA ASF Safety Advisor "GPS Technology," which provides an excellent overview of GPS operations.
- The articles "About GPS" and "GPS Guide for Beginners" on the Garmin website.

locations for DME arcs practice flights

The DME Arcs Practice Flights use approach procedures from other categories of IFR Practice Flights, such as the BRUTE5 departure from Medford, OR (KMFR).

performance standards

These Practice Flights relate to Area of Operation V, "Navigation Systems" in the FAA IFR Practical Test Standards. Task A, "Intercepting and Tracking Navigational Systems and DME arcs," which outlines the following standards:

Objective—To determine that the applicant:

- Exhibits adequate knowledge of the elements related to intercepting and tracking navigational systems and DME arcs.
- Tunes and correctly identifies the navigation facility.
- Sets and correctly orients the course to be intercepted into the course selector or correctly identifies the course on the RMI.
- Intercepts the specified course at a predetermined angle, inbound or outbound from a navigational facility.
- Maintains the airspeed within +/–10 knots, altitude within +/–100 feet, and selected headings within +/–5°.
- Applies proper correction to maintain a course, allowing no more than three-quarter-scale deflection of the CDI or within +/–10° in case of an RMI.
- Determines the aircraft position relative to the navigational facility or from a waypoint in the case of GPS.
- Intercepts a DME arc and maintain that arc within +/–1 nautical mile.
- Recognizes navigational receiver or facility failure, and when required, reports the failure to ATC.

ILS approaches

Use the Practice Flights in this section to observe and practice flying ILS approaches, including:

- Full ILS procedures that require procedure turns
- Approaches joined after following charted transitions
- Approaches that begin from a simulated vector to join the localizer
- ILS approaches that end with a sidestep maneuver to a parallel runway.
- ILS approaches that include a circle-to-land maneuver to land on a runway that is not aligned with the localizer

background information

Before using the ILS Approach Practice Flights, you should be familiar with the following background information:

The sections about the ILS in Chapter 7, Navigation Systems, in the *Instrument Flying Handbook*.

The sections about circling, sidestep, and other approach variations in Chapter 10 of the *Instrument Flying Handbook*.

The sections about the ILS in Chapter 5 of the *Instrument Procedures Handbook*.

"Basic Instrument Flight Patterns" at the end of Chapter 5 of the *Instrument Flying Handbook*.

The information about the ILS in AIM 1-1-9.

The Safety Advisor "Single-Pilot IFR" available in the online library at the AOPA Air Safety Foundation website.

locations for ILS practice flights

The ILS Approach Practice Flights use approaches at Hillsboro, OR (KHIO), Bellingham, WA (KBLI), and other airports in the Pacific Northwest.

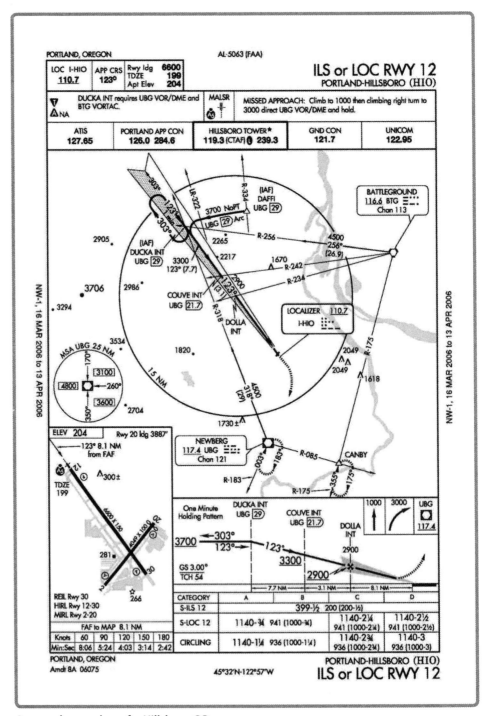

Approach procedures for Hillsboro, OR

performance standards

These Practice Flights relate to Area of Operation VI, "Instrument Approach Procedures" in the FAA IFR Practical Test Standards. Task B, "Precision Approach," outlines the applicable standards, which include determining that the pilot—

- Exhibits adequate knowledge of the precision instrument approach procedures.
- Complies, in a timely manner, with all clearances, instructions, and procedures.
- Advises ATC anytime that the applicant is unable to comply with a clearance.
- Establishes the appropriate airplane configuration and airspeed/V-speed considering turbulence, wind shear, microburst conditions, or other meteorological and operating conditions.
- Completes the aircraft checklist items appropriate to the phase of flight or approach segment, including engine out approach and landing checklists, if appropriate.
- Prior to beginning the final approach segment, maintains the desired altitude +/−100 feet, the desired airspeed within +/−10 knots, the desired heading within +/−10°; and accurately tracks radials, courses, and bearings.
- Selects, tunes, identifies, and monitors the operational status of ground and airplane navigation equipment used for the approach.
- Applies the necessary adjustments to the published DA/DH and visibility criteria for the airplane approach category as required, such as:

  −NOTAMs
  −inoperative airplane and ground navigation equipment
  −inoperative visual aids associated with the landing environment
  −NWS reporting factors and criteria

- Establishes a predetermined rate of descent at the point where the electronic glide slope begins, which approximates that required for the aircraft to follow the glide slope.
- Maintains a stabilized final approach, from the Final Approach Fix to DA/DH allowing no more than three-quarter scale deflection of either the glide slope or localizer indications and maintains the desired airspeed within +/−10 knots.
- A missed approach or transition to a landing shall be initiated at Decision Height.
- Initiates immediately the missed approach when at the DA/DH, and the required visual references for the runway are not unmistakably visible and identifiable.
- Transitions to a normal landing approach only when the aircraft is in a position from which a descent to a landing on the runway can be made at a normal rate of descent using normal maneuvering.
- Maintains localizer and glide slope within three-quarter scale deflection of the indicators during the visual descent from DA/DH to a point over the runway where glide slope must be abandoned to accomplish a normal landing.

localizer approaches

Use the Practice Flights in this section to observe and practice flying Localizer approaches, including:

- Full localizer approaches that include course reversals
- Localizer approaches joined after following charted transitions
- Localizer approaches that begin from a simulated vector to join the localizer
- Localizer-DME approaches with step-down fixes
- Localizer back course approaches
- Localizer approaches that end with a sidestep maneuver to a parallel runway
- Localizer approaches that include a circle-to-land maneuver to land on a runway that is not aligned with the final approach course.

background information

Before using the Localizer Approach Practice Flights, you should be familiar with the following background information:

The sections about the ILS in Chapter 7 "Navigation Systems" of the *Instrument Flying Handbook*.

The sections about circling, sidestep, and other approach variations in the *Instrument Flying Handbook*.

The sections about the ILS in Chapter 5 of the *Instrument Procedures Handbook*.

"Basic Instrument Flight Patterns" at the end of Chapter 5 of the *Instrument Flying Handbook*.

The information about the ILS in AIM 1-1-9.

The Safety Advisor "Single-Pilot IFR" available in the online library at the AOPA Air Safety Foundation website.

locations for localizer approaches practice flights

The Localizer Approach Practice Flights use approaches at Aurora, OR (KUAO), Arlington, WA (KAWO), and other airports in the Pacific Northwest.

performance standards

These Practice Flights apply to Area of Operation VI "Instrument Approach Procedures" in the FAA IFR Practical Test Standards. Task A, "Nonprecision Approaches," outlines the applicable standards. For a list of the most relevant standards, see the Performance Standards section for "VOR Approaches," earlier in this chapter.

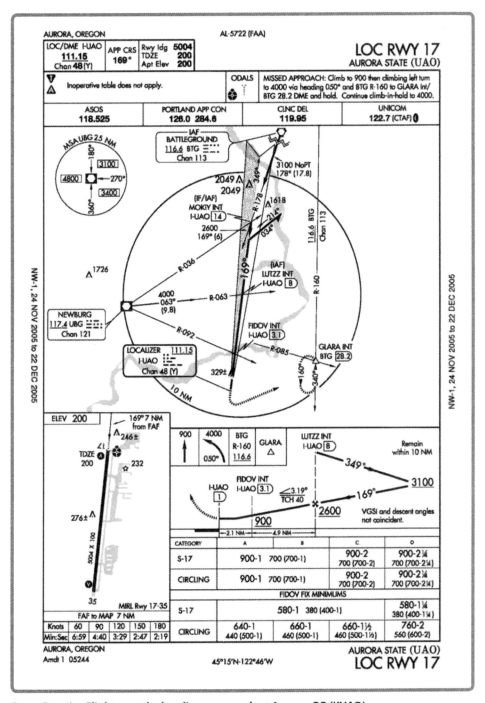

Some Practice Flights use the localizer approach at Aurora, OR (KUAO).

VOR approaches

Use the Practice Flights in this section to observe and practice flying VOR approaches, including:

- Full VOR approaches including procedure turns
- VOR Approaches joined after following charted transitions without a procedure turn
- VOR approaches that include a circle-to-land maneuver to land on a runway that is not aligned with the inbound VOR course
- VOR approaches that incorporate DME and step-downs.

background information

Before using the VOR Approach Practice Flights, you should be familiar with the following background information:

The sections about VOR in Chapter 7 of the *Instrument Flying Handbook*.

The information about DME in Chapter 7 of the *Instrument Flying Handbook*.

The sections about circling, sidestep, and other approach variations in the *Instrument Flying Handbook*.

The sections about VOR approaches in Chapter 5 of the *Instrument Procedures Handbook*.

"Basic Instrument Flight Patterns" at the end of Chapter 5 of the *Instrument Flying Handbook*.

The information about VOR in AIM 1-1-3.

The information about DME in AIM 1-1-7.

The topics "What You Need to Know about VOR" and "Using the Radios" in the Flight Simulator Learning Center.

The Safety Advisor "Single-Pilot IFR" available in the online library at the AOPA Air Safety Foundation website.

locations for VOR approaches practice flights

The VOR Approaches Practice Flights use approaches at Walla Walla, WA (KALW), Aurora, OR (KUAO), Hillsboro, OR (KHIO), Salem, OR, and other airports in the Pacific Northwest.

performance standards

These Practice Flights relate to Area of Operation VI, "Instrument Approach Procedures" in the FAA IFR Practical Test Standards. Task A, "Nonprecision Approach," outlines the applicable standards for all nonprecision approaches, which include determining that the pilot—

- Exhibits adequate knowledge of the elements related to an instrument approach procedure.
- Selects and complies with the appropriate instrument approach procedure to be performed.
- Establishes two-way communications with ATC, as appropriate, to the phase of flight or approach segment, and uses proper communication phraseology and technique.
- Selects, tunes, identifies, and confirms the operational status of navigation equipment to be used for the approach procedure.

- Complies with all clearances issued by ATC or the examiner.
- Recognizes if any flight instrumentation is inaccurate or inoperative, and takes appropriate action.
- Advises ATC or examiner anytime that the aircraft is unable to comply with a clearance.
- Establishes the appropriate aircraft configuration and airspeed considering turbulence and wind shear, and completes the aircraft checklist items appropriate to the phase of the flight.
- Maintains, prior to beginning the final approach segment, altitude within +/–100 feet, heading within +/–10° and allows less than ¾ scale deflection of the CDI or within +/–10° in the case of an RMI, and maintains airspeed within +/–10 knots.
- Applies the necessary adjustments to the published MDA and visibility criteria for the aircraft approach category when required, such as—

a. NOTAMs.
b. inoperative aircraft and ground navigation equipment.
c. inoperative visual aids associated with the landing environment.
d. NWS reporting factors and criteria.

  - Establishes a rate of descent and track that will ensure arrival at the MDA prior to reaching the MAP with the aircraft continuously in a position from which descent to a landing on the intended runway can be made at a normal rate using normal maneuvers.
  - Allows, while on the final approach segment, no more than a three-quarter-scale deflection of the CDI or within 10° in case of an RMI, and maintains airspeed within +/–10 knots of that desired.
  - Maintains the MDA, when reached, within +100 feet, –0 feet to the MAP.
  - Executes the missed approach procedure when the required visual references for the intended runway are not distinctly visible and identifiable at the MAP.

NDB approaches

Use the Practice Flights in this section to observe and practice flying NDB approaches, including:

- Full NDB approaches including procedure turns
- NDB approaches with the beacon located on the airport
- NDB approaches with the beacon along the final approach course
- NDB approaches joined after following charted transitions
- NDB approaches that include a circle-to-land maneuver to land on a runway that is not aligned with the final approach course

background information

Before using the NDB Approach Practice Flights, you should be familiar with the following background information:

The sections about NDB navigation in Chapter 7 of the *Instrument Flying Handbook*.

The sections about NDB approaches in Chapter 5 of the *Instrument Procedures Handbook*.

The detailed information about the NDB/ADF navigation system in AIM 1-1-2.

The topics "Automatic Direction Finder" and "Using the Radios" in the Flight Simulator Learning Center.

The Safety Advisor "Single-Pilot IFR" available in the online library at the AOPA Air Safety Foundation website.

You can find addition useful information about NDB and ADF navigation in Chapter 14 of *The Pilot's Handbook of Aeronautical Knowledge*.

locations for NDB approaches practice flights

The NDB Approaches Practice Flights use approaches at Shelton, WA (KSHN), Hillsboro, OR (KHIO), and other airports in the Pacific Northwest.

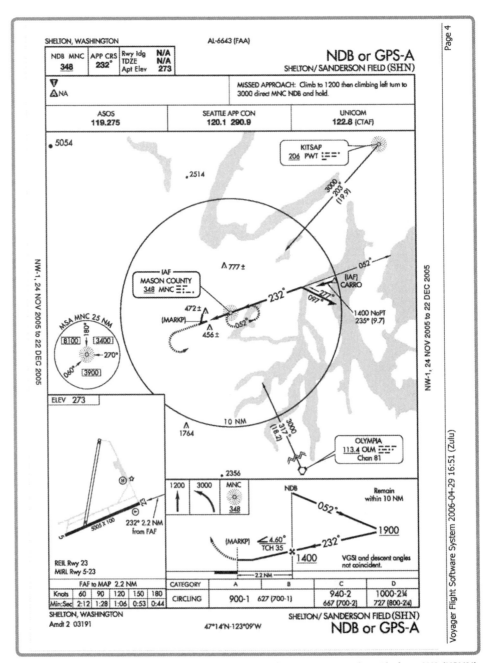

Some of the Practice Flights for NDB procedures use the NDB approach at Shelton, WA (KSHN).

performance standards

These Practice Flights apply to Area of Operation VI "Instrument Approach Procedures" in the FAA IFR Practical Test Standards. Task A, "Nonprecision Approaches," outlines the applicable standards. For a list of the most relevant standards, see the Performance Standards section for "VOR Approaches," earlier in this chapter.

GPS approaches

This area of IFR operations is changing rapidly as new GPS approaches are published, more aircraft are equipped with IFR-approved GPS, and enhancements such as the Wide Area Augmentation System (WAAS) are incorporated into the system. The Practice Flights in this section focus on the following types of procedures:

- Stand-alone (GPS-only) approaches
- GPS overlay approaches that supplant VOR and NDB procedures
- GPS approaches that include a circle-to-land maneuver to a runway that is not aligned with the final approach course

You should use these Practice Flights to help you become generally familiar with GPS-based instrument approaches. However, because operating procedures for the GPS equipment installed in aircraft vary and the GPS, simulated in *Microsoft Flight Simulator* does not reproduce all the features and controls of the actual units, you should take extra care to ensure that you understand the operating procedures for and limitations of the equipment in the aircraft that you fly.

background information

Before using the GPS Approach Practice Flights, you should be familiar with the following background information:

The sections about GPS navigation and approaches in Chapter 7 of the *Instrument Flying Handbook*.

The sections about GPS (RNAV) approaches in the *Instrument Procedures Handbook*.

The detailed information about the GPS navigation system in AIM 1-1-19.

The topic "Using the GPS" in the Flight Simulator Learning Center, which includes details about the GPS in *Microsoft Flight Simulator*.

The Safety Advisors "Single-Pilot IFR" and "GPS Technology" available in the online library at the AOPA Air Safety Foundation website.

The complete Garmin 500 Pilot's Guide available for download from the Garmin website. The Garmin 500 Quick Reference Guide, another handy reference, is also available from the Garmin website. To learn more about GPS, see "About GPS" and "GPS Guide for Beginners" on the Garmin website.

locations for GPS approaches practice flights

The GPS Approaches Practice Flights use approaches at several airports in the Pacific Northwest for which procedures are available in the Flight Simulator database.

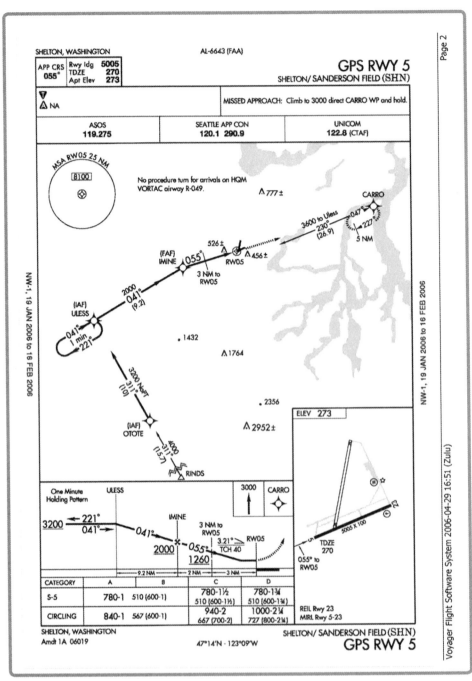

Some of the Practice Flights use the GPS Rwy 5 approach at Shelton, WA (KSHN).

performance standards

These Practice Flights apply to Area of Operation VI "Instrument Approach Procedures" in the FAA IFR Practical Test Standards. Task A, "Nonprecision Approaches," outlines the applicable standards. For a list of the most relevant standards, see the Performance Standards section for "VOR Approaches," earlier in this chapter.

missed approaches

The IFR Practice Flights do not include situations designed specifically to demonstrate and rehearse missed approach procedures. However, you can use any Practice Flight that includes an instrument approach procedure to observe and practice missed approaches.

performance standards

These Practice Flights apply to Area of Operation VI "Instrument Approach Procedures" in the FAA IFR Practical Test Standards. Task C, "Missed Approach," outlines the applicable standards, which include determining that the pilot—

- Exhibits adequate knowledge of the elements related to missed approach procedures associated with standard instrument approaches
- Initiates the missed approach promptly by applying power, establishing a climb attitude, and reducing drag in accordance with the aircraft manufacturer's recommendations
- Reports to ATC beginning the missed approach procedure
- Complies with the published or alternate missed approach procedure
- Advises ATC or examiner anytime that the aircraft is unable to comply with a clearance, restriction, or climb gradient
- Follows the recommended checklist items appropriate to the go-around procedure
- Requests, if appropriate, ATC clearance to the alternate airport, clearance limit, or as directed by the examiner
- Maintains the recommended airspeed within +/−10 knots; heading, course, or bearing within +/−10°; and altitude(s) within +/−100 feet during the missed approach procedure.

circling approaches

The IFR Practice Flights do not include situations designed specifically to demonstrate circle-to-land approach procedures. However, you can use any Practice Flight that includes circling minimums to fly such approaches.

performance standards

These Practice Flights apply to Area of Operation VI "Instrument Approach Procedures" in the FAA IFR Practical Test Standards. Task D, "Circling Approach outlines the applicable standards, which include determining that the pilot—

- Exhibits adequate knowledge of the elements related to a circling approach procedure.
- Selects and complies with the appropriate circling approach procedure considering turbulence and wind shear and considering the maneuvering capabilities of the aircraft.
- Confirms the direction of traffic and adheres to all restrictions and instructions issued by ATC and the examiner.
- Does not exceed the visibility criteria or descend below the appropriate circling altitude until in a position from which a descent to a normal landing can be made.
- Maneuvers the aircraft, after reaching the authorized MDA and maintains that altitude within +100 feet, −0 feet and a flight path that permits a normal landing on a runway. The runway selected must be such that it requires at least a 90° change of direction, from the final approach course, to align the aircraft for landing.

thirteen

sample briefings for IFR practice flights

checklist

☐ VOR briefing

☐ ILS briefing

☐ SID briefing

This chapter includes examples of the preflight briefings that accompany the IFR Practice Flights. As explained in Chapter 12, each Practice Flight includes a detailed briefing in Adobe Reader format (.pdf) that you can print. The complete set of preflight briefings is available on the CD.

> **note**
>
> The IFR charts are reproduced from the Voyager Flight Software System with the permission of Seattle Avionics Software, Inc. and obviously should be used only for simulated flying, not for navigation.

"BruceAir-IFR—ALW-PSC-VOR-Navigation-Briefing"

VOR-navigation

These Practice Flights support learning and rehearsing basic VOR tracking skills.

The Practice Flights begin in the air near Walla Walla, WA (KALW). You are at 4,500 feet in the Cessna 172 or the Beechcraft BE58 Baron with the default instrument panel. The autopilot is ON in HDG and ALT modes, and the simulation is paused.

Practice intercepting and tracking inbound and outbound on radials from VORs in the area and identifying intersections. Use the Practice Flights that include wind to observe and correct for the effects of wind drift.

location

These Practice Flights begin in the air west of Walla Walla, WA (KALW) along V520 between the Walla Walla (ALW) and Pasco (PSC) VORs.

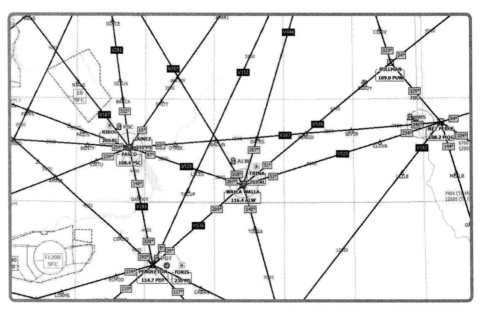

The VOR Navigation Practice Flights begin in the air near Walla Walla, WA (ALW).

background information

Before using the VOR Navigation Practice Flights, you should be familiar with the following background information:

> The sections about VOR navigation, and information about DME, in Chapter 7 of the *Instrument Flying Handbook*.
>
> The information about the VOR navigation system in AIM 1-1-3.
>
> "What You Need to Know about VOR" and "Using the Radios" in the Flight Simulator Learning Center.

If you use the Practice Flights with the Beechcraft BE58 Baron, make sure that you are also familiar with the HSI and RMI instruments, described in Chapters 3 and 7 of the *Instrument Flying Handbook*.

Additional background information about VOR and DME is available in Chapter 14 of the *Pilot's Handbook of Aeronautical Knowledge*.

practice flight file names

flight file name	conditions
BruceAir-IFR--ALW-PSC-VOR-Navigation-C172-01.FLT	Day/VMC/Lt wind
BruceAir-IFR--ALW-PSC-VOR-Navigation-C172-02.FLT	Day/IMC/Wind
BruceAir-IFR--ALW-PSC-VOR-Navigation-C172-03.FLT	Night/VMC/Lt wind
BruceAir-IFR--ALW-PSC-VOR-Navigation-BE58-01.FLT	Day/VMC/Lt wind
BruceAir-IFR--ALW-PSC-VOR-Navigation-BE58-02.FLT	Day/IMC/Lt wind

"BruceAir-IFR—KHIO-ILSorLOC-Rwy12-DAFFI-Briefing"

ILS and localizer approaches

These Practice Flights support learning and rehearsing ILS and localizer approaches. These Practice Flights begin in the air northwest of the Hillsboro, OR airport (KHIO). You are at 4,000 feet in the Cessna 172 or the Beechcraft BE58 Baron with the default instrument panel. The autopilot is ON in HDG and ALT modes, and the simulation is paused.

location

You are north of KHIO on the 334 radial of the UBG VOR near DAFFI intersection. Fly the published transition from DAFFI via the UBG 29 nm DME arc and complete the ILS Rwy 12 approach to KHIO.

starting area on IFR chart

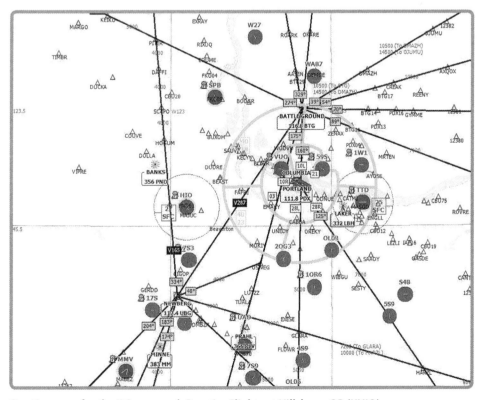

Starting area for the ILS approach Practice Flights at Hillsboro, OR (KHIO).

KHIO ILS or LOC12 chart

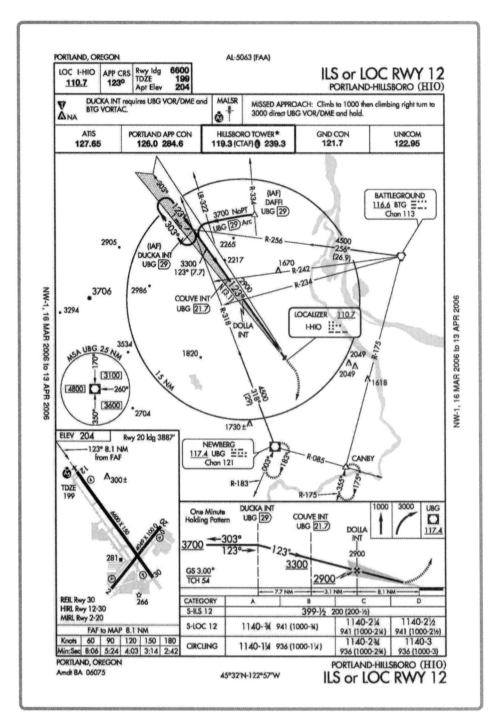

The approach chart for the ILS Rwy 12 approach at Hillsboro, OR (KHIO).

background information

Before using these ILS Approach Practice Flights, you should be familiar with the following background information:

> The sections about the ILS and DME arcs in Chapter 7 of the *Instrument Flying Handbook*.

> The sections about circling, sidestep, and other approach variations in Chapter 10 of the *Instrument Flying Handbook*.

> The sections about the ILS and DME arcs in Chapter 5 of the *Instrument Procedures Handbook*.

> "Basic Instrument Flight Patterns" at the end of Chapter 5 of the *Instrument Flying Handbook*.

> The information about the ILS in AIM 1-1-9.

If you use the Practice Flights with the Beechcraft BE58 Baron, make sure that you are also familiar with the HSI and RMI instruments, described in Chapters 3 and 7 of the *Instrument Flying Handbook*.

practice flight file names

flight file name	conditions
BruceAir-IFR--KHIO-ILSorLOC-Rwy12-DAFFI-C172-01.FLT	Day/VMC/Lt wind
BruceAir-IFR--KHIO-ILSorLOC-Rwy12-DAFFI-C172-02.FLT	Day/IMC/Wind
BruceAir-IFR--KHIO-ILSorLOC-Rwy12-DAFFI-BE58-01.FLT	Day/VMC/Lt wind
BruceAir-IFR--KHIO-ILSorLOC-Rwy12-DAFFI-BE58-02.FLT	Day/IMC/Wind

BruceAir-IFR—KMFR-BRUTE5.BRUTE-Rwy14-Briefing

standard instrument departure procedures (SIDs)

These Practice Flights support learning and rehearsing Standard Instrument Departure Procedures (SIDs).

location

These Practice Flights begin at the Rogue Valley International Airport at Medford, OR (KMFR). You are on the ground in the Cessna 172 or the Beechcraft BE58 Baron with the default instrument panel.

sample
briefings
for IFR
practice
flights

221

KMFR airport diagram

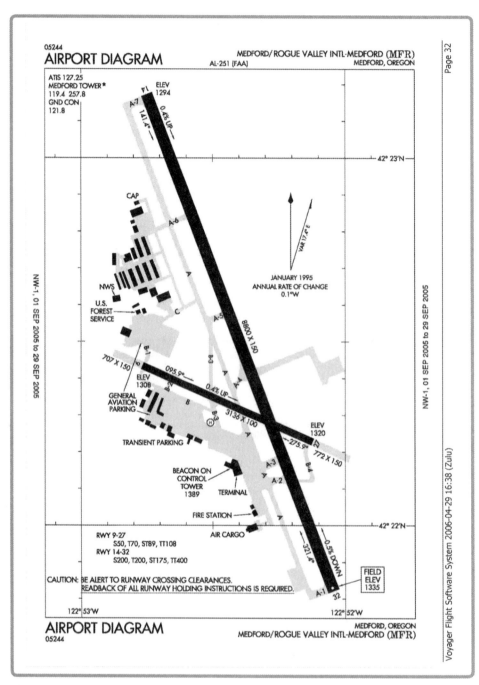

The airport diagram for the Medford/Rouge Valley airport (KMFR).

KMFR BRUTE.BRUTE5 chart

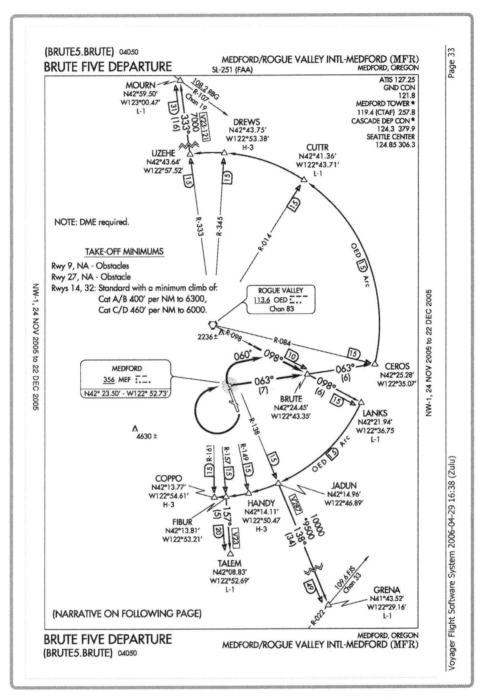

The chart for the challenging BRUTE FIVE departure at Medford, OR (KMFR).

sample
briefings
for IFR
practice
flights

223

BRUTE FIVE DEPARTURE

DEPARTURE ROUTE DESCRIPTION

<u>TAKE-OFF RUNWAY 14</u>: Climbing right turn direct to MEF NDB, depart MEF NDB 063° bearing outbound to BRUTE INT. Thence. . . .

<u>TAKE-OFF RUNWAY 32</u>: Climbing right turn via 060° heading and OED R-098 to BRUTE INT. Thence. . . .

. . . . via (transition) or (assigned route). Maintain 11000 or assigned altitude.

<u>COPPO TRANSITION (BRUTE5.COPPO)</u> : From over BRUTE INT via OED R-098 to LANKS INT. Turn right and proceed via the OED 15 DME Arc to COPPO DME.

<u>CUTTR TRANSITION (BRUTE5.CUTTR)</u> : From over BRUTE INT via MEF NDB bearing 063° to CEROS INT. Turn left and proceed via the OED 15 DME Arc to CUTTR DME.

<u>DREWS TRANSITION (BRUTE5.DREWS)</u> : From over BRUTE INT via MEF NDB bearing 063° to CEROS INT. Turn left and proceed via the OED 15 DME Arc to DREWS DME.

<u>GRENA TRANSITION (BRUTE5.GRENA)</u> : From over BRUTE INT via OED R-098 to LANKS INT. Turn right and proceed via the OED 15 DME Arc to JADUN DME. Then via OED R-138 to GRENA INT.

<u>HANDY TRANSITION (BRUTE5.HANDY)</u> : From over BRUTE INT via OED R-098 to LANKS INT. Turn right and proceed via the OED 15 DME Arc to HANDY DME.

<u>LANKS TRANSITION (BRUTE5.LANKS)</u> : From over BRUTE INT via OED R-098 to LANKS INT.

<u>MOURN TRANSITION (BRUTE5.MOURN)</u> : From over BRUTE INT via MEF NDB bearing 063° to CEROS INT. Turn left and proceed via the OED 15 DME Arc to UZEHE DME. Then via OED R-333 to MOURN INT.

<u>TALEM TRANSITION (BRUTE5.TALEM)</u> : From over BRUTE INT via OED R-098 to LANKS INT. Turn right and proceed via the OED 15 DME Arc to FIBUR DME. Then via OED R-157 to TALEM DME.

<u>UZEHE TRANSITION (BRUTE5.UZEHE)</u> : From over BRUTE INT via MEF NDB bearing 063° to CEROS INT. Turn left and proceed via the OED 15 DME Arc to UZEHE DME.

NW-1, 24 NOV 2005 to 22 DEC 2005

NW-1, 24 NOV 2005 to 22 DEC 2005

Voyager Flight Software System 2006-04-29 16:38 (Zulu)

BRUTE FIVE DEPARTURE
(BRUTE5.BRUTE) 02332

MEDFORD, OREGON
MEDFORD/ROGUE VALLEY INTL-MEDFORD (MFR)

background information

Before using the SID Practice Flights, you should be familiar with the following background information:

The information about departure procedures in Chapter 10 of the *Instrument Flying Handbook* and Chapter 2 of the *Instrument Procedures Handbook*.

The sections about NDB navigation and DME arcs in Chapter 7 of the *Instrument Flying Handbook*.

The sections about DME arcs in Chapter 5 of the *Instrument Procedures Handbook*

Section 5-2, "Departure Procedures," in the AIM.

If you use the Practice Flights with the Beechcraft BE58 Baron, make sure that you are also familiar with the HSI and RMI instruments, described in Chapters 3 and 7 of the *Instrument Flying Handbook*.

starting position as shown on IFR chart

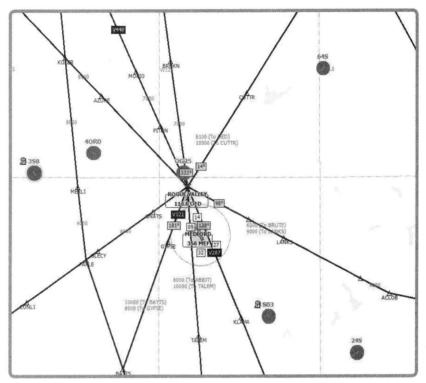

Some of the Departure Procedure Practice Flights begin at the Medford, OR airport (KMFR).

practice flight file names

flight file name	conditions
BruceAir-IFR--KMFR-BRUTE5.BRUTE-Rwy14-C172-01.FLT	Day/VMC/Lt wind
BruceAir-IFR--KMFR-BRUTE5.BRUTE-Rwy14-C172-02.FLT	Day/IMC/Wind
BruceAir-IFR--KMFR-BRUTE5.BRUTE-Rwy14-BE58-01.FLT	Day/VMC/Lt wind
BruceAir-IFR--KMFR-BRUTE5.BRUTE-Rwy14-BE58-02.FLT	Day/IMC/Wind

sample
briefings
for IFR
practice
flights

225

fourteen

creating your own practice flights

checklist

☐ create a flight checklist

☐ choosing an aircraft

☐ starting position

☐ the environment

☐ slew mode

☐ configuring avionics

☐ saving the flight

☐ sharing flights

After you have tried the Practice Flights associated with this book, you may want to create Flights so that you can use Flight Simulator to complement specific lessons or proficiency drills.

Creating a Flight is not complicated. Chapter 7 explained that a Flight is just a set of initial conditions, including:

- An aircraft type and its configuration: position of landing gear and flaps, power setting, lights, etc.
- The aircraft's position: at a specific location on an airport or, if in the air, the aircraft's location, speed, altitude, and heading.
- Configuration of the avionics: frequencies set in the communications and navigation radios, courses selected on navigation controls and indicators, and autopilot status.
- The environment: weather (wind, clouds, visibility, precipitation), season, and time of day.
- The simulation state: paused or running.

Creating a Flight is much like saving any file. Just set up the initial conditions for a flight and then store those conditions, just as you would save a document.

You can find complete details about the Flights feature in the topic "All about Flights" in the Flight Simulator Learning Center."

checklist for creating a flight

Creating a Flight, like flying a real airplane, will go more smoothly if you have a plan. To help you create Flights quickly with a minimum of adjustments, I have provided a "Create a Flight Checklist" in Adobe Reader format (.pdf) that you can copy from the CD and website for this book.

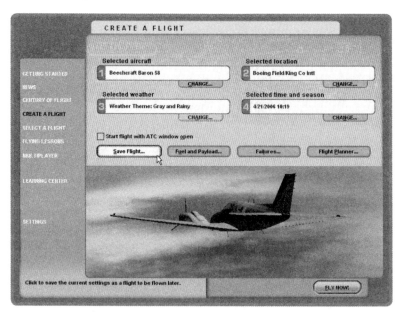

You can set all the starting conditions for a Flight in the Create a Flight dialog box.

flight description

The first step in creating a Flight is to write a basic description. You don't need to compose an essay. Just get an appropriate sectional chart, approach chart, or similar reference, and then jot down essential information about the purpose of the Flight and a brief summary that will appear under **Description** in the **Select Flight** dialog box.

other details

Next, fill in the Create a Flight Checklist to specify the other initial conditions for the Flight as shown in the sample checklist.

With this information at hand, you can quickly set most of the initial conditions for your Flight in the Create a Flight dialog box and efficiently set up the cockpit in the aircraft.

Create a Flight Checklist

Purpose	
Description	
File Name	
Nearest Airport ICAO/Name	
IAP & Transition	

Position	Location	ALT	HDG	KIAS

Date and Time	

Sim Status	Paused	Flying

Weather	Wind	Visibility	Ceiling	Precipitation

Aircraft Config.	MP	RPM	Flaps	Wheels	Cowl

Avionics	Com1	Com2	Nav1	Nav2	ADF	XPND
	/	/	/	/		

DME Source	Nav1	Nav2

Heading Bug/ GPS Waypoint	

Autopilot Config.	Status	HDG	NAV	APR	REV	ALT	±FPM

NAV/GPS Switch	NAV	GPS

Lights	Land	Taxi	Nav	Strobes/ Beacon
				/

Failures	Instruments	Systems	Radio	Engine

© 2006 BruceAir, LLC
www.BruceAir.com

Create a Flight Checklist

choosing an aircraft

Begin the process by choosing an aircraft from the list available under **Selected Aircraft** in the **Create a Flight** dialog box (called **Free Flight** in Flight Simulator X). The list includes all aircraft available to Flight Simulator, including add-on aircraft that you have downloaded and installed from websites or other sources. After you choose an aircraft, click **OK** to return to the **Create a Flight** dialog box for the next steps in the process.

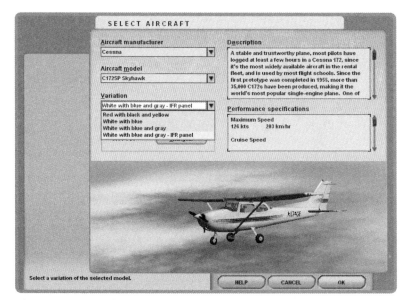

Choose your aircraft.

positioning your aircraft

Next, set the approximate starting position for the Flight. The quickest method, regardless of whether you want to start a Flight on the ground or in the air, is first to use the controls under **Selected Location** to place the aircraft at an airport. Even if you want a Flight to begin in the air, this step places the aircraft in the general area of an approach or other procedure that you want to fly. You can use the Map later to place the aircraft at or near a specific navaid, intersection, IAF, or other location.

For example, to begin positioning the aircraft for a Flight to practice an instrument approach at Bremerton, WA (KPWT), in the **Create a Flight** dialog box, click **Change** under **Selected Location**. Then, in the **Airport ID** box, type **KPWT**. Note that you should use the complete official ICAO or FAA abbreviation for airports. If you don't know the official ID for an airport, you can use the search tools in the dialog box to filter the list of nearly 24,000 airports in Flight Simulator.

After placing the aircraft at an airport, click OK to return to the **Create a Flight** dialog box.

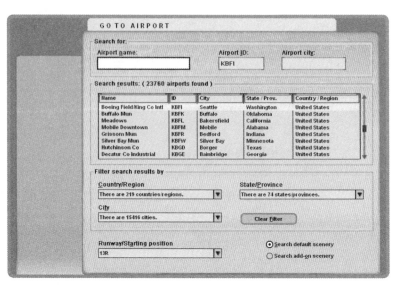

Choose a location to start from.

setting the environment

The next two steps set the weather and the time and season for the Flight.

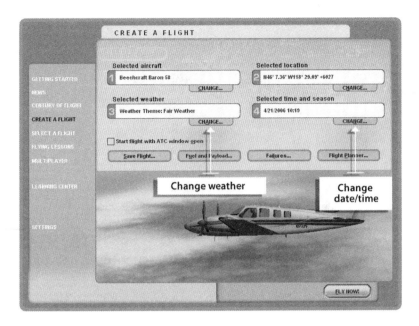

In the **Create a Flight** dialog box, under **Selected Weather**, click **Change**.

In the **Weather** dialog box, you can quickly set general conditions by applying a weather theme, such as "Clear skies" or "Fogged in." To set specific weather that reflects minimums for a particular approach or to generate winds of a specified strength and direction for such exercises as tracking VOR radials and holding, click **User-Defined Weather**, and the use the options in the **Advanced Weather** dialog box.

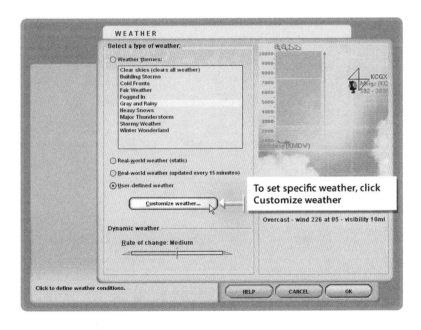

You can learn all about customizing the weather in Flight Simulator in the topic "Introduction to Flight Simulator Weather" in the Learning Center.

creating your own practice flights

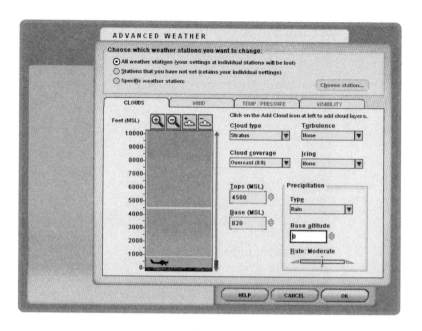

I recommend saving a Flight with specific weather (either a weather theme or user-defined weather), even if later you want to use current conditions based on the Real-World Weather feature in Flight Simulator. Setting specific weather reduces the size of the .wx file associated with the Flight. Also, if a future user of a Flight isn't connected to the Internet, the Flight will begin with the specified conditions. You can change the weather for any Flight and import Real-World Weather at any time after loading a Flight.

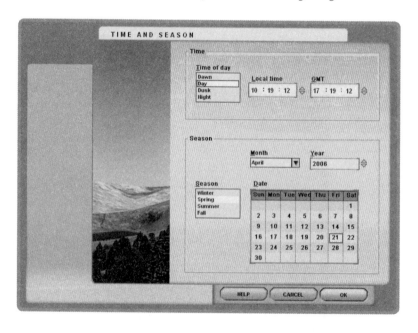

After specifying the weather for the Flight, return to the **Create a Flight** dialog box, and click **Change** under **Selected Time and Season**. Use the controls in the **Time and Season** dialog box to set the time of day and date for the Flight.

After setting the environment and time of day, click the **Fly Now!** button to return to Flight Simulator. The aircraft will be placed on the ground at the airport you specified.

setting a precise starting location with the map

To set a precise location for a Flight that begins in the air, display the **Map** by choosing the Map command on the **World** menu or by clicking the map icon on the instrument panel.

With the Map displayed, you can drag the airplane symbol to a precise location relative to an airport, navaid, airway, or other feature. Use the icons at the top of the Map to select which items appear on the display. You can zoom in and out and scroll the Map by clicking the edges of the chart. If you drag the aircraft symbol to an airport, the snap-to feature places the airplane on a runway.

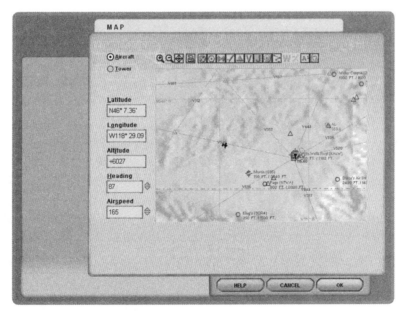

Drag to place the airplane on the map for starting location.

After placing the airplane on the Map, for a Flight that begins in the air, fill in the altitude, heading, and speed boxes to set the aircraft's initial configuration.

For more information about the Map in Flight Simulator, see the topic "Using the Map" in the Learning Center.

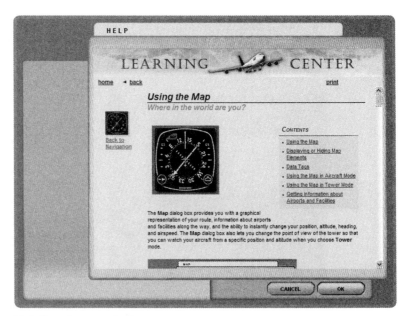

Help for using map features.

It's a good idea to place the aircraft near, but not precisely on, a point that defines an IAF, the beginning of a transition, or other significant point on an instrument procedure. For example, leaving distance from an IAF gives you time to stabilize Flight Simulator before you save the Flight. A little maneuvering room also helps pilots who later use the Flight to orient themselves before they need to turn, change altitudes, intercept a course, or adjust the avionics.

To close the Map and return to the simulation, click OK. Immediately press the P key or, on the World menu, choose Pause Flight. This process freezes the simulation so that you can finish setting up the Flight.

tweaking position with slew mode

After dragging the aircraft to a place on the Map, you may need to fine-tune its position relative to a VOR radial or other reference. If you use a joystick with Flight Simulator, you can make such adjustments with Slew Mode (slewing is cumbersome with a yoke).

To switch Flight Simulator into and out of Slew Mode, press the **Y** key or, on the **Aircraft** menu, click Slew Mode. While **Slew Mode** is active, moving the joystick translates the aircraft's position. For example, moving the joystick to the left slides the aircraft left; pushing the joystick forward slides the aircraft forward; and so on.

To increase altitude slowly while in Slew Mode, press either the **Q** or **F3** key; to decrease altitude, press the **A** key. Holding either of those keys increases the rate of altitude change.

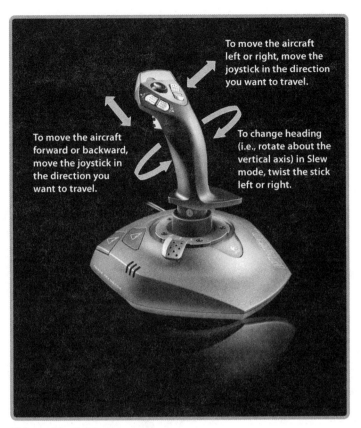

To move the aircraft left or right, move the joystick in the direction you want to travel.

To move the aircraft forward or backward, move the joystick in the direction you want to travel.

To change heading (i.e., rotate about the vertical axis) in Slew mode, twist the stick left or right.

Using a joystick in slew mode.

using slew mode

For more information about using Slew Mode, see the topic "Positioning Your Aircraft" in the Learning Center.

basic slew actions	joystick movement or key
Switch into or out of Slew Mode	Press **Y** key
Move forward	Push joystick forward or press **Num Pad 8**
Move backward	Pull joystick back or press **Num Pad 2**
Move left	Move joystick left or press **Num Pad 4**
Move right	Move joystick right or press **Num Pad 6**
Freeze slewing	Press **Num Pad 5**
Move up	Press **Q** or **F3**
Move down	Press **A**
Rotate nose to new heading	Twist joystick left or right or press **Num Pad 1** or **Num Pad 3**
Set heading north and attitude straight-and-level	Press **SPACEBAR**

configuring the avionics

After positioning the aircraft with the map and pausing Flight Simulator, you can complete configuration of avionics and aircraft systems. The process resembles doing a cockpit flow check.

First, set up the communications and navigation radios, the course selectors, and the autopilot. It is best to select HDG and ALT modes on the autopilot and set the heading bug and the altitude to match the settings you specified on the Map.

You can set up the avionics in the avionics stack, which opens in a separate window.

setting aircraft configuration and power

Next, set other aircraft controls such as the flaps, landing gear, and lights, as appropriate for the initial conditions required for the Flight.

Finally, press the **P** key or, on the **World** menu, choose **Pause Flight** to resume the simulation so that you can set power and stabilize the aircraft. As soon as Flight Simulator resumes "flying," use the controls on a joystick or yoke or the mouse pointer to set the throttle position and other power controls.

You can use the configuration tables described in Chapter 8, "Flying the Aircraft Used in the Practice Flights," to help you quickly configure the aircraft for common situations.

Give the aircraft a few moments to stabilize at an appropriate airspeed and wait for the autopilot to guide the airplane to the specified heading and altitude.

Next, press **P** to pause the Flight in a stable configuration.

Engine controls window for the Baron

saving the flight

You are now ready to save the set of initial conditions you've set up as a Flight. On the **Flights** menu, choose **Save Flight**. In the **Title of Flight** box, type a file name for your Flight. I suggest using a file naming convention that concisely describes the purpose of the Flight. Following a consistent naming pattern makes it easy to find specific Flights in the Select a Flight dialog box, especially if you share your Flights with others. For examples, see "Title and File Naming Conventions for Practice Flights," in Chapter 7, "About the Practice Flights."

In the **Description** box, type basic information about the Flight to help others understand the purpose and location of the scenario. (Tip: You can write the text for the description in a word processing application and then cut-and-paste it directly into the **Description** box.) Keep the description short and simple. If you want to include more details and operating hints, you can create a preflight briefing for a Flight. For more information about that process, see "Creating a Briefing" in the Learning Center.

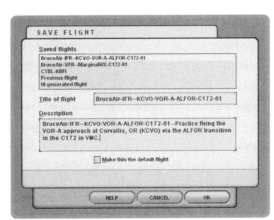

Save your flight.

using your flights

To use Flights that you create, follow the guidelines in Chapter 7, "About the Practice Flights." For more information, see the topic "All About Flights" in the Learning Center.

Note that each time you load a Flight that begins in the air you should give the simulation a few seconds to stabilize. You may also need to adjust power settings, depending on where the last person left the throttle levers, and you may also need to make other adjustments, such as setting the altimeter by pressing the **B** key.

sharing flights

Flights and their associated files are easy to bundle into small packages that you can share with a USB memory stick, burn onto a CD, attach to an email message, or offer for download from a website.

Just send all of the files associated with a Flight (.flt, .wx, .pln, and .htm files, as appropriate) from the **My Documents\Flight Simulator Files** folder on your hard drive. The recipients can copy the files to the same directory on their hard drives.

Because all of the files for a category of Flights are located in a single folder, it is easy to share an entire category with other Flight Simulator users—just provide a copy of the category folder and all of its contents. To install a folder containing a category of Flights, copy the folder into the **Flight Simulator 9\Flights** or **Flight Simulator 10\Flights** folder. The new category and its flights appear in the **Select a Flight** dialog box.

For more information about creating categories of Flights and sharing Flights, see the topic "All about Flights" in the Learning Center.

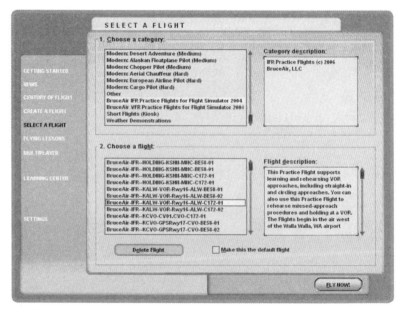

Flights that you create appear in the Select a Flight dialog box.

index

A

advanced "training features" **89**
airport operations **157, 182**
airspace **175**
AOPA Air Safety Foundation **13, 21**
arrivals (STARs) **197**
ASA **12**
ATC **59, 114**
ATC 6 **10, 20**
autopilot **58, 78**
aviation books **12**

B

basic attitude instrument flying **189**
Basic Aviation Training Device (B-ATD) **47**
basic flying skills **154, 180**
basic instrument maneuvers **170**
BATD **46**
BE58 Baron **30**
Bernoulli's Principle **22**
Boeing 747, **30**
briefings **179, 217**

C

capturing a screen **57**
CD contents **147**
Cessna 172, **30, 35**
charts **146**
circling approaches **215**
classroom **56**
cockpit flows **25**
cockpits **118**
commercial pilot practice flights **154, 177**
complex aircraft **177**
computer requirements **5**

conventional cockpit

conventional cockpit **121**
Crew Resource Management (CRM) **94**
crosswind approach and landing **160**
crosswind takeoff and climb **160**
CTAF **25**
customizing Flight Simulator **87**

D

database **44**
dead reckoning **167**
departure procedures **195**
develop "the numbers" for aircraft **61**
diversions **167**
DME arcs **204**

E

emergencies **176**
en route navigation **196**
evaluating **115**

F

FAA approval **45**
FAA aviation handbooks **142**
FAA references **145**
fixation **40**
flight analysis **83**
flight instructor's utility **92**
flight management computers **21**
flight model **29**
flights **80**
Flight Safety International **18**
Flight Simulator 2004 **67, 90, 92, 97**
Flight Simulator X **43, 67, 91, 97, 110, 229**

flight training devices 31, 32, 46
flying "the numbers" 118
force-feedback 31
FTD 28, 94

G

"Glass" panel 15
Garmin 295, 84
Garmin 500, 204
Garmin G1000 113, 123, 133
Garmin G1000 glass cockpit 100
Garmin GPS 84
go-around 162
GPS 84
GPS approaches 213
GPS navigation 168

H

holding procedures 199

I

IFR practice flights 187
IFR training panels 90
ILS 219
ILS approaches 205
IMAX® film 31
instant replay 83
integrated flight instruction 41

J

Jeppesen charts 12, 146
joystick or yoke 46, 70
joysticks, yokes, throttles, rudder pedals 6

K

kneeboard 69

L

landings 159
learning curve 45
learning plateaus 52, 55
Link Trainer 20
localizer approaches 208
logging time 45, 47
lost procedures 168

M

map 82
marginal weather 173
Microsoft 1
Microsoft Flight Simulator 2004 2

Microsoft Flight Simulator X 2
missed approaches 215
mouse 72, 100
Multifunction Display 101

N

NACO charts 145
NASA Langley Research Center 33
navigation 166
NDB approaches 211
NDB navigation 168, 193
negative transfer 27, 41
night flying 172

O

Obstacle Departure Procedure 195
online 146

P

PCATD 46
perception 31
peripheral vision 31
personal computer based aviation training
  devices 47
pilotage 167
Piper Cub 15
Piper Warrior 35
practice flights 103, 227
preflight briefings 108
Primary Flight Display 101
private pilot practice flights 154
procedure turns 202
projectors 56

R

radios 156
RAM 5
realism 11, 29, 34

S

shared aircraft 94
sharing flights 238
short-field 159
side slips 22
Simkits 17
slew mode 234
slow flight 164
soft-field 159
stalls 164
stalls, spins 33
standard instrument departure 221
Standard Instrument Departure procedures 195
State Dependent Learning 29
steep turns 163

swiss army knife 24
systems and equipment malfunctions 177
systems failures 86

T

"the numbers" for flying the Beechcraft BE 58
  Baron 134
"the numbers" for flying the Skyhawk 124
takeoffs 159
traffic patterns 157, 182
training handbooks 11
trim 119

U

U.S. Air Force 37
U.S. Navy 18
unusual flight attitudes 171

V

VFR flying 52
VFR practice flights 151
video (graphics) card 5
videos 83
view modes 75
view system 98
VOR approaches 210
VOR navigation 52, 168, 191, 218

W

weather 85
web links 141
wrap-around displays 31
www.BruceAir.com 3

X

X-Plane 46

Y

Yogi Berra 22, 58
yoke, rudder pedals, power levers 27